茶道经
YĪM YÁŊ
CANON OF TEA

汉英对照

by Li Hui

Revised by ZHu Yongsheng

上海大学出版社　复旦大學出版社

图书在版编目(CIP)数据

茶道经：汉英对照 / 李辉著. —上海：上海大学出版社,2023.7

ISBN 978-7-5671-4718-8

Ⅰ. ①茶…　Ⅱ. ①李…　Ⅲ. ①茶文化-中国-汉、英　Ⅳ. ①TS971.21

中国国家版本馆 CIP 数据核字(2023)第 090186 号

本书版权授予方为 復旦大學出版社 Fudan University Press

责任编辑　许家骏　戴文沁
封面设计　汪　超
技术编辑　金　鑫　钱宇坤

茶道经(汉英对照)

李　辉　著

上海大学出版社出版发行

(上海市上大路 99 号　邮政编码 200444)

(https://www.shupress.cn　发行热线 021-66135112)

出版人　戴骏豪

*

南京展望文化发展有限公司排版

上海光扬印务有限公司印刷　各地新华书店经销

开本 890mm×1240mm　1/32　印张 5　字数 116 千

2023 年 7 月第 1 版　2023 年 7 月第 1 次印刷

ISBN 978-7-5671-4718-8/TS·21　定价　118.00 元

To memorialize the four founders of tea study at Fudan University

WU, Jue Nong
(1897-1989)

CHEN, Chuan
(1908-1999)

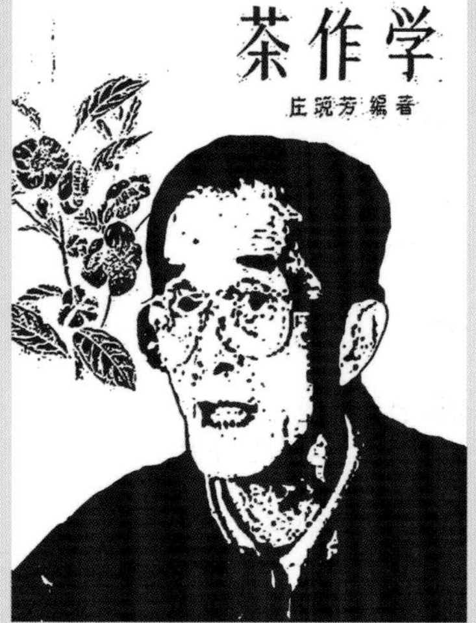

ZHUANG, Wan Fang
(1908-1996)

WANG, Ze Nong
(1907-1999)

FOREWORD

Tea is one of the most popular drinks of the world. People all know that tea originated in China, and there are more tea cultivars and products (all the six types of tea) in China. However, for a long history, only green tea and black tea were introduced into Europe, and most of the so-called black teas in Europe are actually red teas except for those in Russia and Turkey. Few Europeans know that there are tea types other than green tea and black tea, or that the tea types are created by different processing techniques rather than plantation. Although there are so many famous names of Chinese teas prevalent in Europe, such as Bohea, Oolong, and Keemun, the diversity of Chinese teas is too strange, let alone the various techniques and medical functions of the teas.

The reason for European people to love tea must be the same as that for Chinese people. That is drinking teas make one comfortable from the tongue to the whole body. There are definitely medical functions in the teas. In Chinese legends, tea was found by the Hot Emperor, Shennong, as the best medicine around 6,000 years ago. In the 14th century, red tea (now is mistakenly called black tea) was introduced into Europe by the Portuguese navigator to replace the rhubarb (*Rheum* sp.) to cure indigestion. It soon became popular in the aristocracy where too many meats were consumed. In the present time, most Europeans eat meat every day and a cup of red tea after meal will help a lot for digestion. That is why Europeans consume so large amount of tea.

In contrast, tea drinking became an art instead of medicine in China and Japan in the recent one thousand years. As a result, most of the Chinese people drink very few teas and are unaware of the medical functions of the teas. However, unless the medical functions are taken into consideration, the qualification of the teas cannot be scientific. In the last decades, many strange teas were invented with a pretty good smell but their ingredients are harmful to health. There must be scientific standards for tea qualification. That is why I wrote this book after a decade of field investigation and laboratory research.

In this book, I explained the fundamental knowledge about the Chinese teas with the traditional Chinese philosophy beside science. In Chinese philosophy, the nature of everything can be classified into Yim and Yaŋ. For example, male is Yaŋ while female is Yim, life is Yaŋ while death is Yim, the sun is Yaŋ while the moon is Yim, and the sky is Yaŋ while the earth is Yim. For the teas, those biochemical reactions are carried out before the leaf cells are killed are Yaŋ teas (living teas), while those reacted after the leaf cells are killed are Yim teas (dead teas). The energy of reactions might come from three origins, i.e., the sky, the human, and the earth, which are called the Three Talents. That is why there are six types of Chinese teas, green, blue, red, white, black, and yellow teas. They go respectively into six meridians of body fluid indicated by Chinese traditional medicine, and therefore, nourish the organs through which these meridians go. For example, red tea is piled on the ground to ferment, which is made by the earth energy, before the cells are killed by heating. Then, it is the Earth Yaŋ (Little Yaŋ) tea, that goes into the Little Yaŋ meridian which nourishes the gallbladder and cures indigestion.

Everything about tea starts from this 2×3 scheme. This is the tea philosophy. We may call it teaism, while it is totally different from the teaism of Japan. Without the tea philosophy, the tea science cannot work, especially the medicine aspect of the teas. In this book, I want to introduce all the six types of the teas and their medical relevance to the world. I hope more and more people will know how to benefit from the teas. By drinking a certain type of tea, one can keep the related organ healthy. For a large number of disorders, we just need a pot of correct tea. A healthy body for everyone around the world is my best wish, and also the wish of this book.

Li, Hui

Fall of 2019

During yellow tea making in Guizhou.

Terminology revision

In this book, we (my research team) established a new system of tea science and philosophy, which is based on the human body meridian system of the Traditional Chinese Medicine. The new system was proved by many experiments and cohort exploration. Interestingly, although the meridian system was proved by the research, some recent hypotheses of the meridians were rejected. For example, one of the twelve meridian channels is Xinbao channel, and people suppose that Xinbao is the pericardium according to the terminology of western medicine. However, in our experiment, the Xinbao is clearly thymus, which is located tightly with the pericardium and has all the expected functions of the Xinbao channel. Therefore, we changed the name "pericardium meridian" into "thymus channel". The same reason is for the revision of "SAN JIAO meridian" into "three-gland channel" which has been though to related to no organs while through our experiments this meridian is related to three endocrine glands, i. e., pituitary, thyroid, and adrenal glands. The Small Intestine meridian and Large Intestine meridian are focused onto the duodenum and colon with experimental evidences. Duodenum (and jejunum) was the exact concept of Xiaochang (literal small intestine) defined in the original Chinese medicine scripture, ***The Yellow Emperor's Canon of Medicine***, in which Xiaochang and Huichang (ileum) were differentiated, and therefore, Xiaochang is not exactly small intestine of modern anatomy. By the way, the rectum of large intestine is actually controlled by the Gladder channel. The revisions were listed in the following table.

In this book	Old name	Chinese name
Grand Yaŋ	Taiyang	太阳
Middle Yaŋ	Yangming	阳明
Little Yaŋ	Shaoyang	少阳
Grand Yim	Taiyin	太阴
Middle Yim	Jueyin	厥阴
Little Yim	Shaoyin	少阴
Hand · Grand Yaŋ · Duodenum Channel	Hand Taiyang Small Intestine Channel/meridian	手太阳小肠经
Foot · Grand Yaŋ · Bladder Channel	Foot Taiyang Urinary Bladder Channel/meridian	足太阳膀胱经
Hand · Middle Yaŋ · Colon Channel	Hand Yangming Large Intestine Channel/meridian	手阳明大肠经
Foot · Middle Yaŋ · Stomach Channel	Foot Yangming Stomach Channel/meridian	足阳明胃经
Hand · Little Yaŋ · Three-gland Channel	Hand Shaoyang Triple Warmer/Burner/Energizer/Sanjiao Channel/meridian	手少阳三焦经

续 表

In this book	Old name	Chinese name
Foot · Little Yaŋ · Gallbladder Channel	Foot Shaoyang Gallbladder Channel/meridian	足少阳胆经
Hand · Grand Yim · Lung Channel	Hand Taiyin Lung Channel/meridian	手太阴肺经
Foot · Grand Yim · Spleen Channel	Foot Taiyin Spleen Channel/meridian	足太阴脾经
Hand · Middle Yim · Thymus Channel	Hand Jueyin Pericardium Channel/meridian	手厥阴心包经
Foot · Middle Yim · Liver Channel	Foot Jueyin Liver Channel/meridian	足厥阴肝经
Hand · Little Yim · Heart Channel	Hand Shaoyin Heart Channel/meridian	手少阴心经
Foot · Little Yim · Kidney Channel	Foot Shaoyin Kidney Channel/meridian	足少阴肾经
Blue tea	Oolong tea(most of the blue teas)	青茶
Red tea	Black tea	红茶
Black tea	Dark tea	黑茶

For the spelling of Yim-Yaŋ instead of Yin-Yang, we considered the original concepts of this couple of terms, Yim as going inside and Yaŋ as coming outside. When one pronounces [-m], the breath goes inside, while [-ŋ] will let breath come outside. That was why traditional Chinese pronounced this couple of words as Yim and Yaŋ.

However, in the recent several hundred years, Mandarin Chinese has lost the -m ending. Here we keep using traditional Chinese to express the original concepts.

目　录
CONTENTS

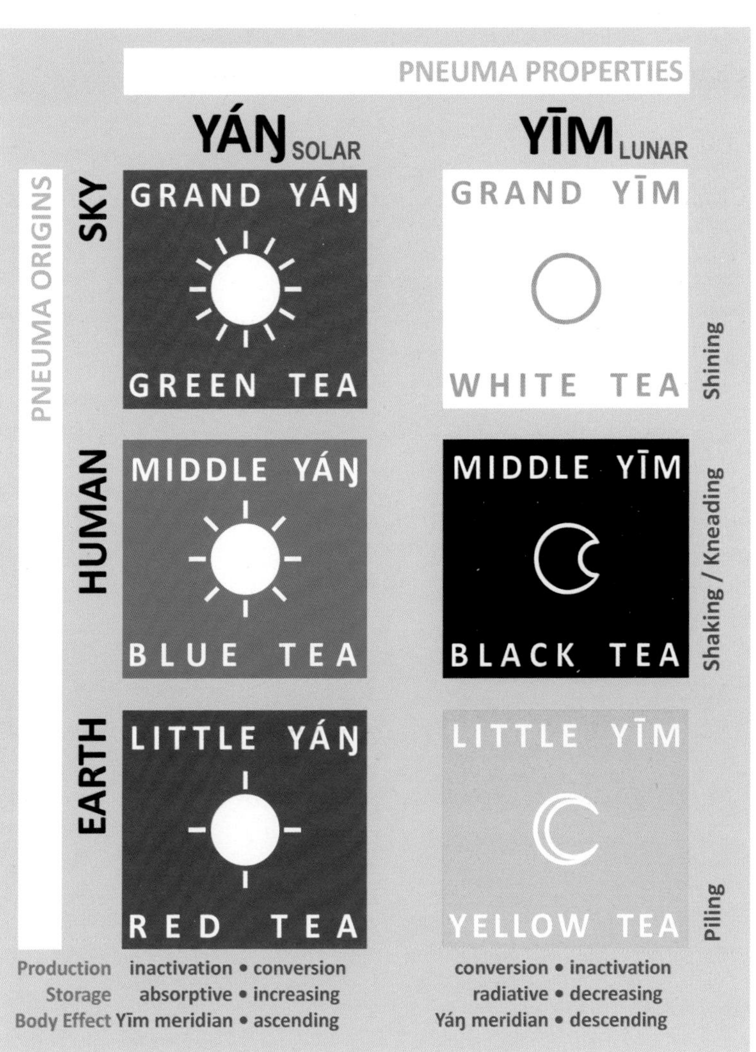

Figure 1. Philosophy of tea classification according to production procedures.

一之类

Chapter 1 CLASSIFICATION

一阴一阳之谓道。茶道者,茶气之阴阳也。阴阳和谐,则天下乐进。观天下之茶,因采制之术相异而分阴阳。未杀青而成分已化、茶气已成者谓之阳茶,杀青而后转化成分、积累茶气者谓之阴茶。阴阳之茶,各应于天地人三才之造,盖有六类。

The balance of Yim (Lunar) and Yaŋ (Solar) elements is called Tao (principle, the basic rule of evolvement). Tao of Tea (Tea Philosophy) is the principle of Yim-Yaŋ classification of the teas. With the balance of Yim and Yaŋ, the teas will be well accepted by all kinds of people. The teas in the world can be classified into Yim teas and Yaŋ teas according to their respective pneumas growing up during fermentations. Those teas with the nutrients and the pneumas created before the enzymes in the tea leaves are inactivated are called the Yaŋ teas (Solar teas); while those with the nutrients and the pneumas created after the enzymes are inactivated are called the Yim teas (Lunar teas). During fermentation, three types of major influences, i.e., the sky, the human, and the earth, will be exercised on Yim and Yaŋ teas, and therefore, the six types of tea are made.

得天之气、杀青不化者谓之绿茶,气属太阳,多咖啡因而利尿提神;

With the vital essence of the sky, green teas inactivated directly without any fermentation, keeps the original pneumas accumulated from sunshine during growing. Their pneumas are Grand Yaŋ. Grand Yaŋ pneumas are healthy for refreshing and urination, because caffeine, which has been proved to have the above-mentioned functions, is kept in the teas.

因人之力、摇青发酵者谓之青茶,气属阳明,多丹宁酸而清肠排毒;

With the hard efforts of the human, blue teas are shaken to ferment. The pneumas of blue teas are Middle Yaŋ. Their pneumas are healthy for stomach and colon because the aromatic acids are created in the blue teas with these functions.

就地之厚,渥堆发酵者谓之红茶,气属少阳,多茶多胺而利胆养颜;

With the great virtue of the earth, red teas are piled up on the ground to ferment. Their pneumas are Little Yaŋ. Their pneumas are healthy for endocrine and gallbladder while amines and amides in red teas have those functions.

就地而烘青闷黄者谓之黄茶,气属少阴,多黄酮类而活血洗肾;

With the influence of the earth, yellow teas are inactivated by baking and subsequently wrapped up by moist bags to oxidize on the ground. Their pneumas are Little Yim. Their pneumas are healthy for cardiovascular and kidneys, and the flavones in yellow teas have those functions.

因人而炒青种曲者谓之黑茶,气属厥阴,多芳香苷而疏肝安神;

With the influence of the human, black teas are inactivated by pan-firing and fermented with fungi. Their pneumas are Middle Yim. Their pneumas are healthy for liver and emotion. The aromatic glycosides in black teas have those functions.

得天而晒青陈化者谓之白茶,气属太阴,多白茶酯而健脾润肺。

With the influence of the sky, white teas are dried in the sun to be inactive and keep basking to oxidize. Their pneumas are Grand Yim. They are healthy for immunity and breath. The esters in white teas have those functions.

阳者发散,阴者收敛。故阳茶日久则气失,不复可饮;阴茶则存之愈久,其气愈浓。

Yaŋ pneumas keep dissipating, while Yim pneumas keep absorbing. Therefore, Yaŋ teas will lose pneuma after a period and will not be edible. On the other side, the longer the Yim teas keep, the better they will be, because more and more pneumas are absorbed.

Table 1 Philosophy of tea classification according to pneuma properties

Character Origin	YIM	YAŊ	conversion
SKY	GRAND YIM WHITE TEA	GRAND YAŊ GREEN TEA	shining
HUMAN	MIDDLE YIM BLACK TEA	MIDDLE YAŊ BLUE TEA	shaking or kneading
EARTH	LITTLE YIM YELLOW TEA	LITTLE YAŊ RED TEA	piling
production	inactivation · conversion	conversion · inactivation	
storage	absorptive · increasing	radiative · decreasing	
extraction	inactive · boiling	active · brewing	
body effect	Yim meridian · ascending	Yaŋ meridian · descending	

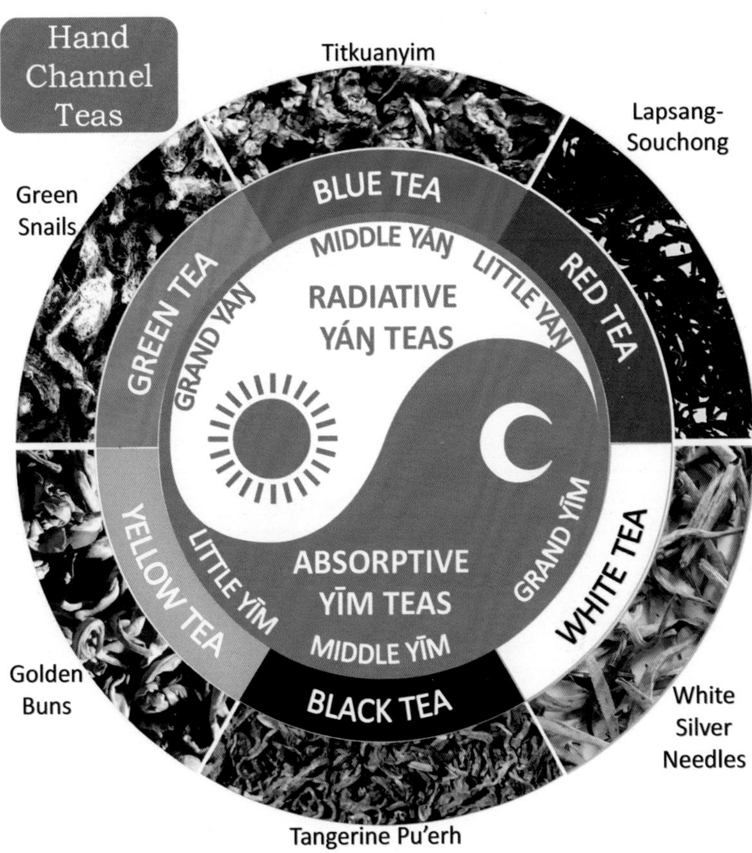

Figure 2a. Classical teas with pneumas going to six channels connected with hands.

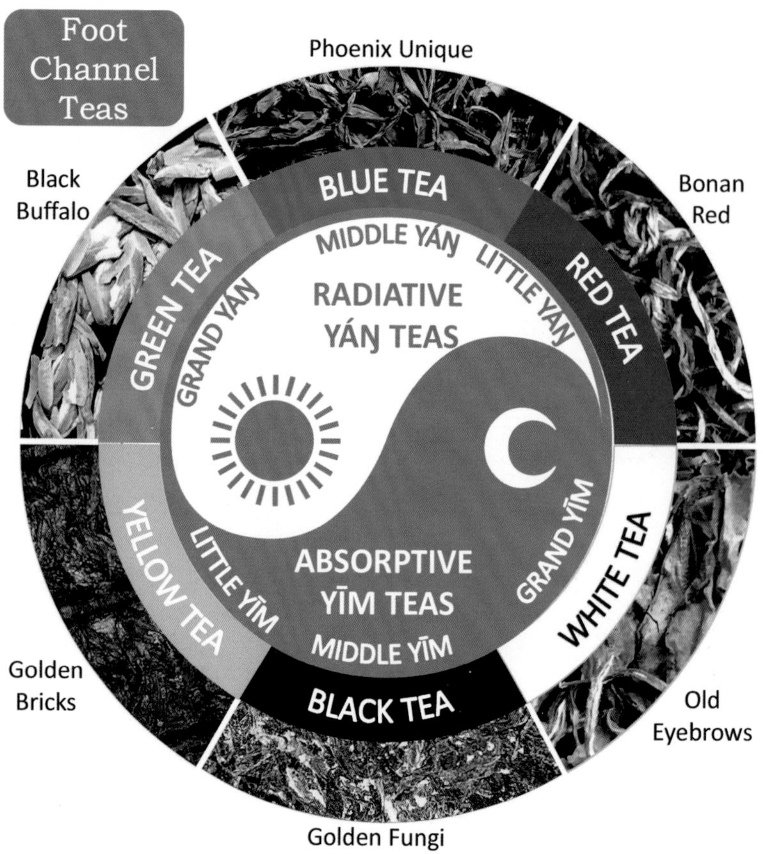

Figure 2b. Classical teas with pneumas going to six channels connected with feet.

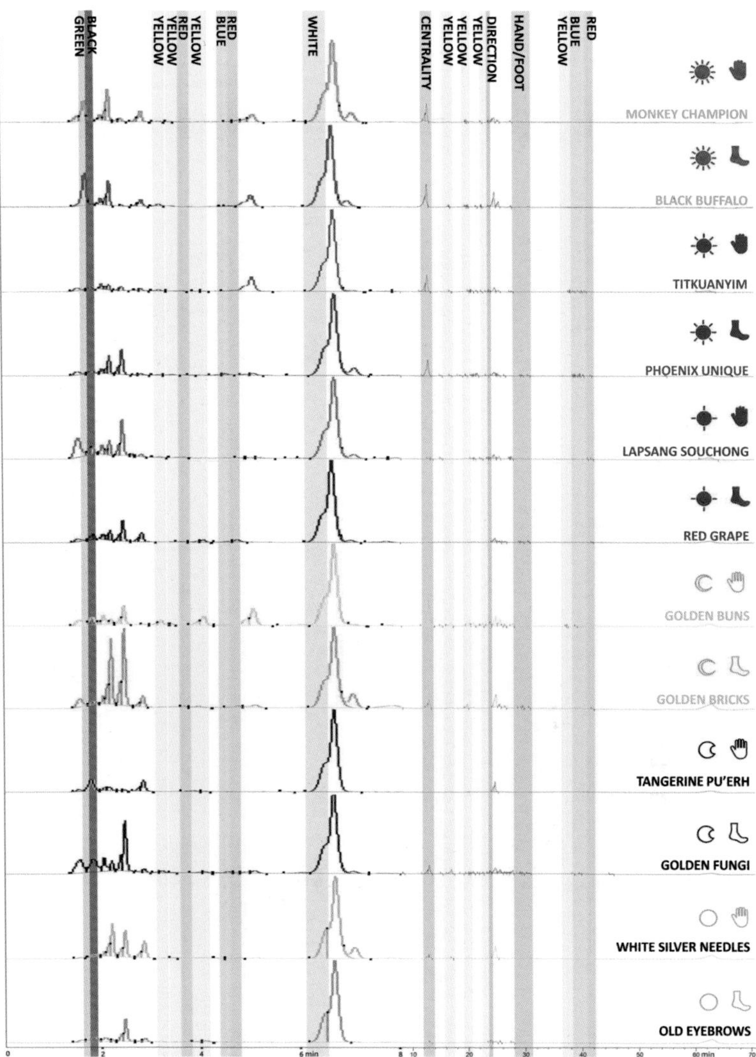

Figure 3. RP-electrophoresis based on molecular polarity isolates unique components for each type of tea, indicating that six types of tea contain totally different chemical compounds and therefore have different health relevance.

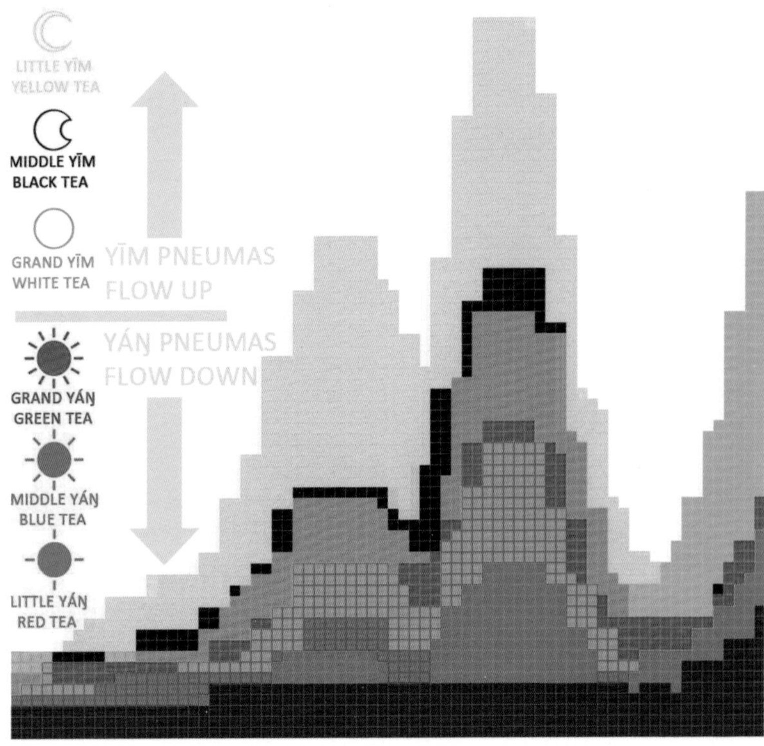

Figure 4. Peaks released by electrophoresis at around 24 minutes demonstrate a divergence of Yīm and Yáŋ teas, which might be the compounds that decided the directions of the body fluid.

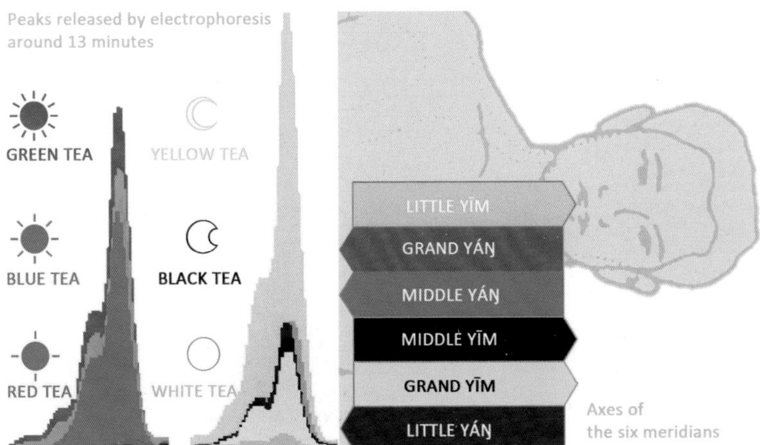

Figure 5. The six teas go to six meridians of body fluid. The meridians line vertically through body. Peaks released by electrophoresis at around 13 minutes demonstrate the order of meridian axes from the body center to the margins.

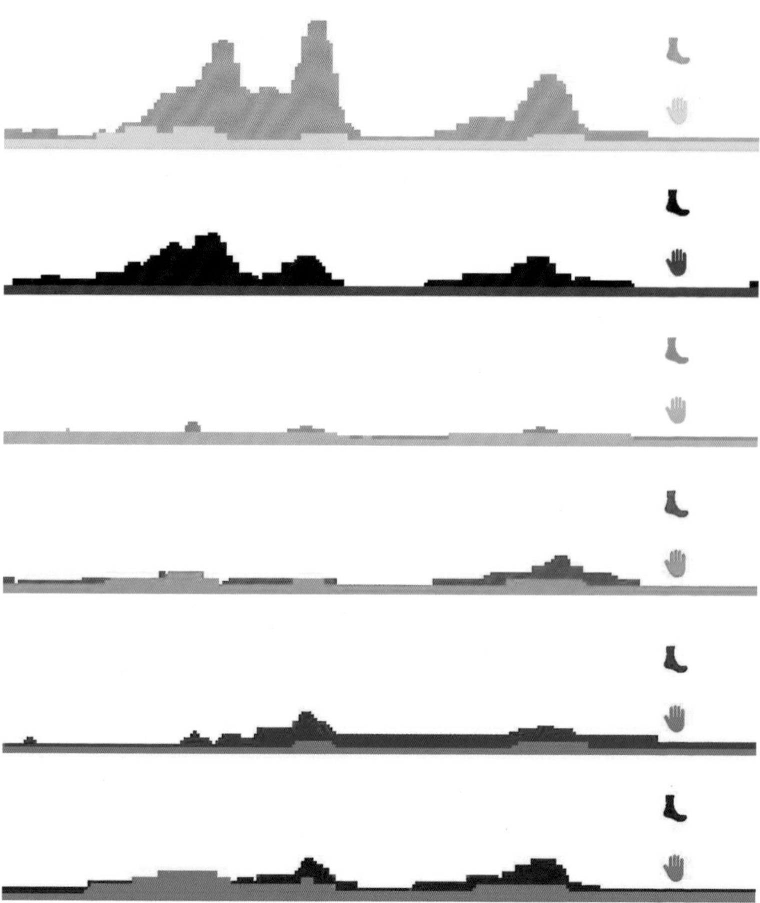

Figure 6. Peaks released by electrophoresis at around 30 minutes demonstrate a divergence of Hand and Foot channels the teas go through.

二之气

Chapter 2 **PNEUMA**

气者,体魄之资也。述其性则有阴阳之分,述其源则有天地人之类。

Pneumas are raw materials of soul. There are two kinds of pneumas (Yim and Yaŋ) according to their performances, and three types of pneumas according to their resources.

人传之气曰宗气,入于奇脉;

Those from human resources are called Ancestral Pneumas and go into the Extra Meridians of the human body.

地产之气曰营气,入于正脉;

Those from earth resources are called Nutritional Pneumas and go into the Normal Meridians of the human body.

天降之气曰卫气,入于肌肤。

Those from sky resources are called Protectional Pneumas and go into the muscles.

茶气者,地产之气也,营气之类也。营气入于正脉而补于营魄。正脉有六,而茶之营气因有六也。《道德经》曰"戴营魄抱一",此养生之本也。故善饮茶者,得营气之利,必强其营魄,而健其身心。

Pneumas of the teas are from earth resources, and therefore, are Nutritional Pneumas.

Nutritional Pneumas go into normal meridians, and nourish the nutritional souls (normal souls related to organs). Chinese traditional medicine believes that souls include three brain souls and seven body souls. Among the seven body souls, six are nutritional and one is ancestral. As there are six normal meridians, the tea pneumas also include six types. "*Tao Te Ching*" said "Taking the nutritional souls into one center", which is the most important principle of health regimen. Those who are good at drinking teas can benefit from Nutritional Pneumas, strengthen their nutritional souls, and make their body healthy.

然则气乃魂魄之属,非物也,人无感官可辨者,何以知之?曰通感也。人有情绪之变,非食饴药而觉甘苦,此通感也。气之入体,必感心神,而通相似之觉,故而饮佳茶有回甘。

However, pneumas are kinds of soul, not of real material, which cannot be felt by human sense organs. Then, how can people observe them? That is synaesthete. When one's emotion fluctuates, he will feel sweet or bitter, although he eats nothing sweet or bitter. That is called synaesthete. When pneumas are taken into the body, certain souls react and the body feels like eating something that causes the same feeling. That is why people feel the reflective sweet after drinking good teas.

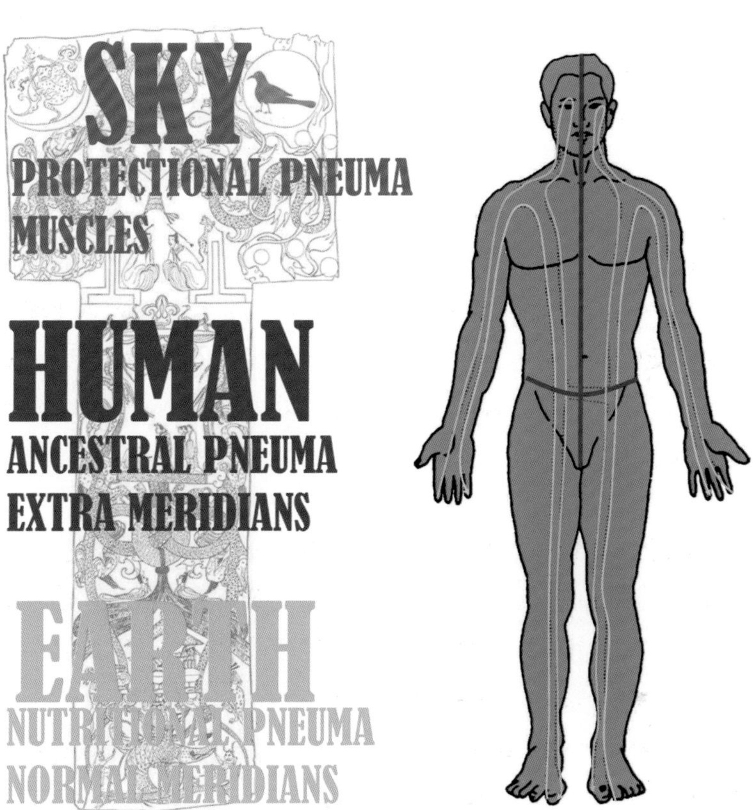

Figure 7. There are three kinds of pneumas judging by their resources of sky, human, and earth, which are called the Three Capacities in Chinese philosophy. These three pneumas go to different parts of the human body.

如黄茶通心肾，枸杞、山楂亦通心肾，故而黄茶回甘若枸杞、山楂；

Just like the pneumas of yellow teas, the wolfberries (*Lycium barbarum*) or haws (*Crataegus pinnatifida*) go into the heart and kidney channels (the Little Yim meridian). Therefore, the yellow teas taste like wolfberries or haws.

绿茶通小肠提神而有胎菊之香，通膀胱利尿而有蚕豆之香；

Likewise, those green teas that go into the duodenum channel and refresh the drinker taste like chrysanthemum (*Chrysanthemum morifolium*), and those that go into the bladder channel taste like broad beans (*Vicia faba*).

红茶通三焦者有巧克力乃至桂圆之香，利胆者有金钱草乃至葡萄干之香；

Those red teas that go into the three-endocrine-gland channel taste like chocolate or longan (*Dimocarpus longan*), and those that go into the gallbladder channel taste like desmodiums (*Glechoma longituba*) or even grapes (*Vitis vinifera*).

黑茶疏肝者气若参芪，通心包者气若陈皮；

Those black teas that go into the liver channel taste like asiabell (*Codonopsis pilosula*) or astragalus (*Astragalus propinquus*), and those that go into the thymus channel taste like dried tangerine (*Citrus reticulata* cv. Chachiensis) peel.

青茶清肠者回甘若木莲花蜜,舒胃者回甘若金桂香皮;

Those blue teas that go into the colon channel taste like magnolia (*Magnolia liliiflora*), sophora flower (*Sophora japonica*) or lychee (*Litchi chinensis*), and those that go into stomach channel taste like cassia (*Cinnamomum cassia*) or osmanthus (*Osmanthus fragrans*).

白茶健脾者有枣香,润肺者有梨甘。

Those white teas that go into the spleen channel taste like jujubes (*Drypetes congestiflora*), and those teas that go into the lung channel taste like pears (*Pyrus bretschneideri*).

Figure 8. **Teas with different pneumas have different flavors which might be related to their major components or the synaesthete caused by the flow of pneumas in the body.**

三之植

Chapter 3 **PLANTATION**

种茶之道,乃集天地间之卫气,转为营气于芽叶中。茶气转化或阴或阳,故种茶须阴阳均和为上。

The principle of tea plantation is collecting protectional pneumas from the space and transforming them into nutritional pneumas of the leaves. Pneumas of the leaves can be Yim or Yaŋ, which requires the plantation to keep the balance of Yim and Yaŋ.

所谓"阳崖阴林",阴阳二气和也。二气充盈均和之所,方可得上等之茶。山高为崖,愈上愈阳;岚浓蔽林,故亦成阴。故山高云密之处出好茶。此天之阴阳均和也。

The saying of "Yaŋ cliff and Yim woods" means a kind of balance between Yaŋ and Yim pneumas.

Where there is a balance of sufficient two pneumas, there grow the best teas. Cliffs are on high mountains, and the higher a cliff is, the more Yaŋ the pneumas are. When the thick haze covers the woods, the Yim pneuma occurs. That is why the best teas grow well on high mountains with thick haze. This is the balance between Yim and Yaŋ in the sky.

茶之"上者生烂石",石为阳,若生黄土,其质阴,则下矣;亦须傍溪涧,水为阴。此地之阴阳均和也。

The best teas grow in broken stones. Stones are Yaŋ in contrast to the loess. Those teas growing in the loess will not be good enough because the loess is Yim. The best teas also should grow nearby a stream. Streams are Yim properly. This is the balance between Yim and Yaŋ in the earth.

植茶者当为“精行俭德”之君子,其性阳而有益;若骄奢淫逸,则其性阴而不当。有言采茶者以少女为上,取其阴当其位而和于阳也。此人之阴阳均和也。

Tea planters should be noble-hearted gentlemen, whose pneuma is Yaŋ and helpful to the teas. If the teas are planted by the person lacking moral fiber, the bad Yim pneumas will be soiled. It says the best teas should be picked by little girls, whose good Yim pneumas can balance the Yaŋ pneumas of the planters. This is the balance between Yim and Yaŋ in the human.

故茶之成株,“艺而不实,植而罕茂”,不可过阳也。

With the balance of Yim and Yaŋ, the tea trees blossom but have few fruits, and grow up but look thin, avoiding excessive bias to Yaŋ.

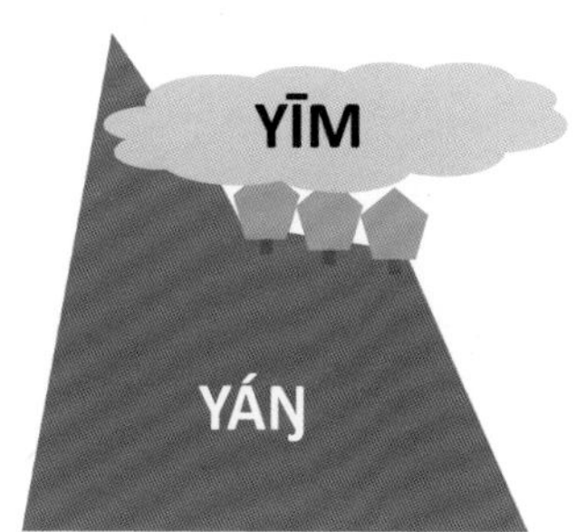

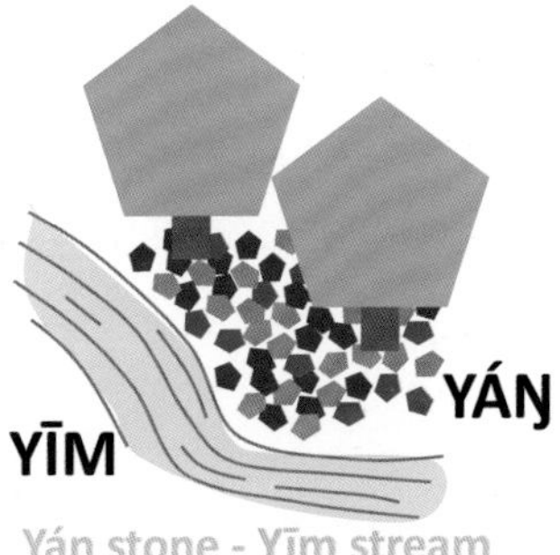

Figure 9. The principles of the plantation of the tea trees are the balance of Yim and Yaŋ in all aspects of the sky, the earth, and the human.

四之造

Chapter 4 **FERMENTATION**

制茶之道,阴阳转换之功也。茶之初出,得日光之精华,其气太阳。因工艺而化其气,顺而为少阴,逆而为阳明,少阴可转厥阴而后太阴,阳明可转少阳。

The principle of tea fermentation is the transformation of Yim and Yaŋ pneumas.

When tea leaves grow up, they obtain pneumas from the sunshine. Thus, the pneumas are Grand Yaŋ, which means the sun. By different techniques, the Grand Yaŋ pneumas transform forwards into Little Yim, or backwards into Middle Yaŋ. Little Yim can be transformed into Middle Yim and then Grand Yim. Middle Yaŋ can be transformed into Little Yaŋ.

阴阳之气,道之为言也。若言其法,乃所含成分之反应也。六茶之中,主要成分有异,为酚,为酮,为酸,为胺,为酯,为苷。其物愈纯,其茶愈善,故必控转化之机,不可不及或过之。

The description of Yim and Yaŋ pneumas is a statement of Tao (philosophy principle). By its scientific principle, it refers to the various biochemical reactions and reaction products. When making six kinds of tea, the products with medical functions are different, mainly polyphenols, flavones, tannic acids, amides, esters, or glycosides. Unprecise technique will produce teas with various complicated products which might cause disorders of the human body

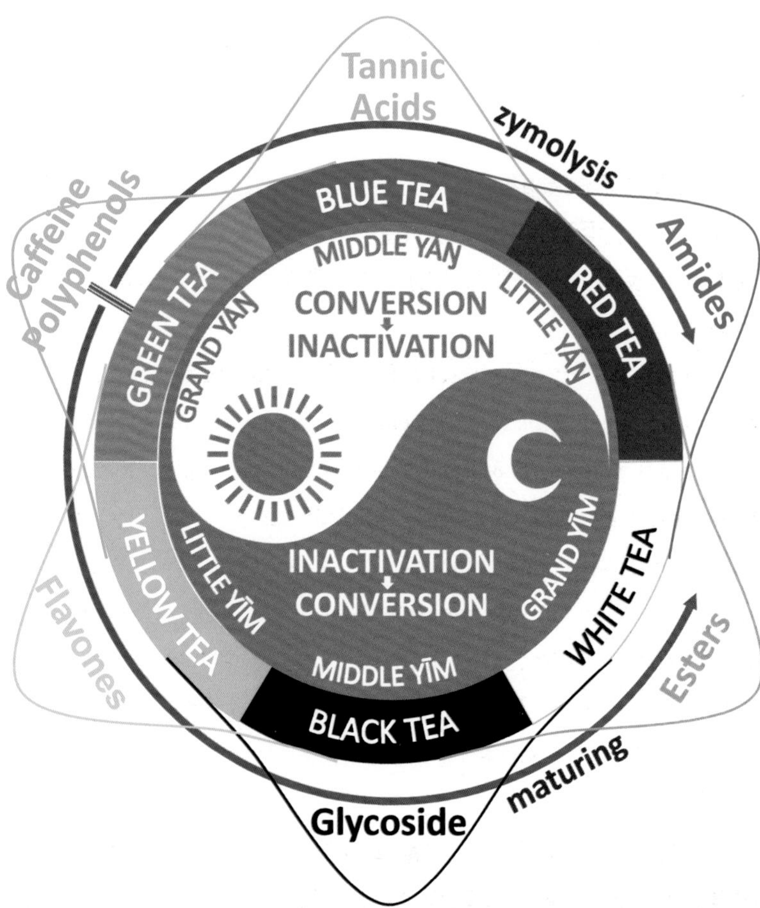

Figure 10. Six types of tea contain different major components. To produce these components, either zymolysis or maturing is performed during tea fermentations. It is most important to seize the opportunity of reaction peaks to make best teas.

after drinking. Therefore, the purer these products are, the better the teas are made. One must seize the opportunity of reaction completion to make best teas. Either before or after the completions, the teas turn bad.

世言黄茶微发酵、青茶半发酵、红茶全发酵,甚谬也! 或微或半,皆中途而废,转化未致,岂可得好茶。实则六茶各有工艺,使其化学反应各异,而得产物不同,殊途而归异也,非同途而半道取者。

Since 1980s, people started to say that yellow tea is lightly fermented, blue tea is half fermented, and red tea is fully fermented. That is not the truth. When teas are lightly or half fermented, the biochemical reactions are stopped before completions, and the teas cannot be good enough without completeness. Actually, the six kinds of tea are made by different techniques. Different reactions produce different components. That means the biochemical reactions have different ways of reactive direction and different ends, not the same way with different extents.

究其工艺,原理有二。以酶催化者乃发酵,可谓之酶化,其酶或自鲜叶,或自菌曲。无酶而氧化者可谓之陈化。绿茶以火灭酶而干封,不使茶多酚及咖啡因流失,自日照所生天然之物尚存,因而存太阳气。

There are two principles of techniques. Fermentation with enzymes is called zymolysis. Thc enzymes may come from leaves or fungi. Oxidation without enzymes is called maturing. The enzymes of

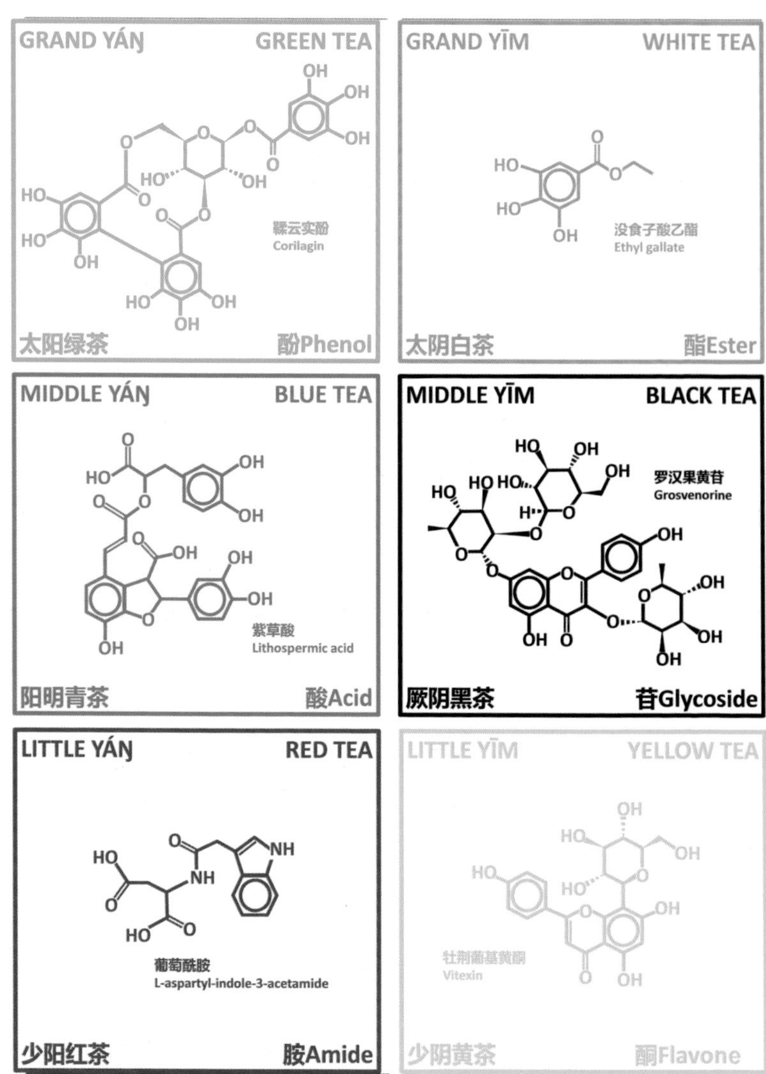

Figure 11. Typical examples of the major functional compounds of the six types of tea.

green teas are inactivated by heating, whereas polyphenols and caffeine are encased. Those produced when growing in sunshine are still kept, which stores the Grand Yaŋ pneuma.

酶化之茶有青红二种。青茶摇青使酶溢出,发酵成酸,人力所用,得阳明气。红茶就地湿热渥堆发酵生胺,地力所蓄,而得少阳气。

Two kinds of tea are made by zymolysis, i.e., blue teas and red teas. When making blue teas, we shake the leaves to break the cells and release the enzymes. These enzymes, together with oxygen, fermented the tannins into tannic acids. When shaken by the human, the pneumas in the blue teas are Middle Yaŋ. When making red teas, we roll up the leaves to break them, and pile the tea buds up on the ground to ferment the amino acids into amides. As the red teas are transformed on the ground, the pneumas are Little Yaŋ.

陈化之茶有黄白二种。黄茶烘青灭酶,使其就地包闷氧化而成酮类,因成少阴气。白茶晒青灭酶,长年缓慢氧化而生酸,继而因太阳热照成酯,渐生太阴气。

Another two kinds of tea are made by maturing, i.e., yellow teas and white teas. The yellow teas are baked to inactivate the enzymes, and wrapped in bags on the ground to be oxidized to product flavones. Thus, the Little Yim pneumas are created. The white teas are basked in the sun to inactivate the enzymes, and are oxidized in a long period to produce fulvic acids, and then heated by sunshine into esters. Thus, the Grand Yim pneumas are created.

Figure 12. Samples of the key techniques in the protocols of tea productions.

黑茶初者炒青灭酶,揉捻氧化,继有阴凉种曲,缓慢酶化兼陈化而得芳香苷,因得厥阴气。

The black teas are fried to inactivate the enzymes at the beginning, and then are pounded to be oxidized. Then they will be dried in the sun and piled up to grow the fungi. Slow zymolysis and maturing create aromatic glycosides, which makes the Middle Yim.

各茶所化成分有异,则所用底物亦须有异。春茶芽叶未展之时,以氮生长为主,多茶氨酸,则宜化胺而作红茶。待生三四叶,则转以碳生长为主,可化酸而为青茶,或化糖而为黑茶。绿茶以一芽或加一叶,得茶多酚之最。白茶可芽可叶,皆有所化。维黄茶以一叶为绝,登黄酮醇之峰时也。

The products of every kind of tea biochemical reactions are different, and therefore, they need different substrates. In early spring, before the buds and leaves open, the growth of the teas are mainly nitridation, and the amino acids are accumulated, which are the proper substrates to transform into amides to make the red teas. When the tea buds open to three or four leaves, the growth changes to carbonation and tannins are accumulated, which can be transformed into tannic acids of the blue teas or into glycosides of the black teas. The green teas use the first buds or the first buds plus the first leaves, thus, they keep the most polyphenols. One can use either buds or leaves to make white teas, while only the first leaves can be used to make yellow teas because they contain the highest concentrations of flavonols.

Table 2 Major procedures of the six types of tea

		growth	water loss	zymolysis	inactivation	cell breaking	maturing	fermentation	shaping	dehydration
YÁŊ	**Green**	shining	airing		heating					baking
	Blue		airing	shaking	heating	kneading			rolling	baking
	Red		airing-rolling	piling	heating					baking
YĪM	**Yellow**		airing		heating	kneading	wrapping		rolling	baking
	Black		airing		heating		pounding	shining-piling	cake making	airing
	White		languishing		shining		shining			baking

五之存

Chapter 5 **PRESERVATION**

藏茶之道,在阴阳相生也。阳茶阴封,阴茶阳养。绿青红三种阳茶,可密封阴冷保存不使气散。白黑黄三种阴茶,则宜室温蕴养。

Tea preservation of the teas must follow the principle of reciprocal transformation between Yim and Yaŋ. Namely, the Yaŋ teas must be sealed by Yim pneumas, and the Yim teas must be nourished by Yaŋ pneumas. The three Yaŋ teas, Green, Blue, and Red, can be sealed and keep cool to avoid the dissipation of the pneumas. On the other hand, the three Yim teas including Yellow, Black, and White teas must be kept in room temperature.

白茶太阴,好太阳气,为天之阳气,即日月之光也。宜装罐日煦常温保存,不时开罐透气,以利其深度氧化。不宜紧压为饼疏绝氧气。

The white teas are Grand Yim, which need to be nourished by Grand Yaŋ pneumas coming from the sky. That is in the sunshine. Therefore, the white teas can be kept in cans and dried in the sunshine (<54 ℃). It is necessary to open the cans occasionally to let oxygen in. Deep oxidation is required for white teas. It is not good to press white teas into hard cakes because that will drive oxygen away from the tea leaves.

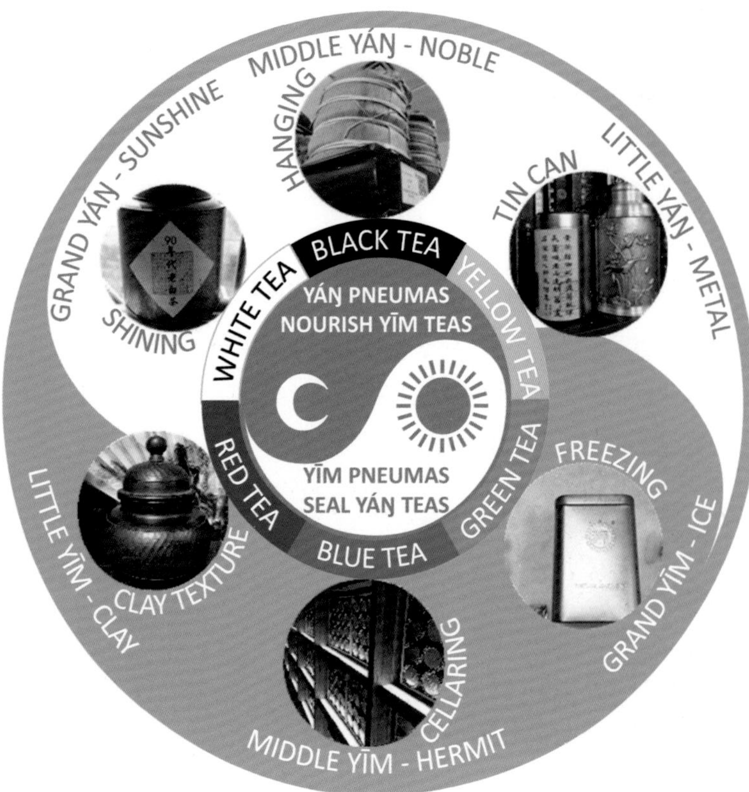

Figure 13. The principle of tea preservation is the reciprocal transformation between Yīm and Yáŋ.

黑茶厥阴,好阳明气,为人之阳气,即君子之气也。人体呼出之气,清者上升,浊者下沉。故黑茶须置室内高处通风保存,高者阳也,可收阳明清气;落地则必收污浊之气。

The black teas are Middle Yim, which need to be nourished by Middle Yaŋ pneumas coming from the humans. That is in the breath of gentlemen. For the human breath, those clean breath is lighter than air and will rise up, while foul breath is heavier and will settle down. Thus, the black teas should be kept high and clean in the room. High is Yaŋ. The black teas can obtain the clean Middle Yaŋ pneumas when kept high. If put on floor, the black teas will obtain the foul pneumas and turn bad.

黄茶少阴,好少阳气,为地之阳气,即金玉之质也。故须密封于锡罐中,以金属养其少阴。锡者,五金之至阳也。

The yellow teas are Little Yim, which need to be nourished by Little Yaŋ pneumas coming from the earth. That is in the metals. The best way of yellow tea preservation is to seal them in tin cans. Tin is the most Yaŋ metal.

绿茶太阳,必须密封冰冻,至阴方可愈岁。寒者太阴气也。

The green teas are Grand Yaŋ, which need to be sealed by Grand Yim pneumas. That is why green teas need to be sealed in cans and frozen. Coldness is Grand Yim.

青茶阳明,乌龙、铁观音全发酵者无须冰冻,同于岩茶,可锡封阴凉置之,愈低愈善,窖藏最佳。低者厥阴气也。

The blue teas are Middle Yaŋ, which need to be sealed by Middle Yim pneumas. That is the low side reclusive pneuma. Therefore, the blue teas should be sealed in cans and kept in a possible lowest position. The fully fermented blue teas including Oolongs, Titkuanyim, and the rock teas do not need to be frozen, but just sealed in tin cans and kept in a cool and low place.

红茶少阳,密封阴藏即可,以水底泥陶封之为佳。泥者少阴气也。

The red teas are Little Yaŋ which need to be sealed by Little Yim pneumas. Mud contains Little Yim pneumas, and therefore, the red teas should be sealed in clay textures and kept cool.

九为至阳之数,阳气耗散,以九年为限。九年之后不复其气,或转为阴茶,或腐朽不堪。太阳绿茶三年即不可继,偶有气纯者,存之得当而转少阴黄茶,或更转厥阴黑茶。阳明青茶,气纯者多转厥阴黑茶。少阳红茶,有转太阴白茶者,或转厥阴黑茶者。

Nine is the largest number of Yaŋ. That is why the Yaŋ teas can only be preserved for nine years. The Grand Yaŋ green teas can only be kept for three years at most. A few green teas will transform into Little Yim yellow teas or even Middle Yim black teas. The Middle Yaŋ pneumas of blue teas or the Little Yaŋ pneumas of red teas cannot be kept for more than nine years. After nine years, these two kinds of tea will rot or transform into Middle Yim black teas or even Grand Yim white teas.

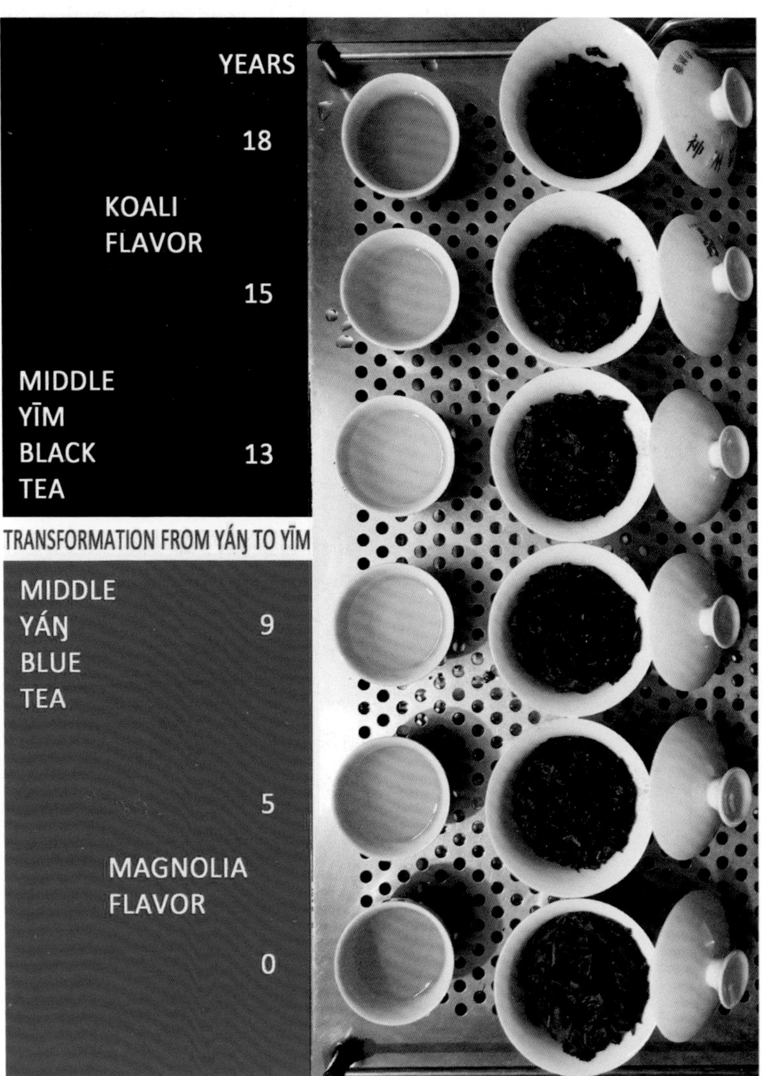

Figure 14. The blue tea Titkuanyim, made by Mr. WANG Qingwen from Fengfeng Mountain of Fujian, turns from Yáŋ tea to Yīm tea gradually in nine years.

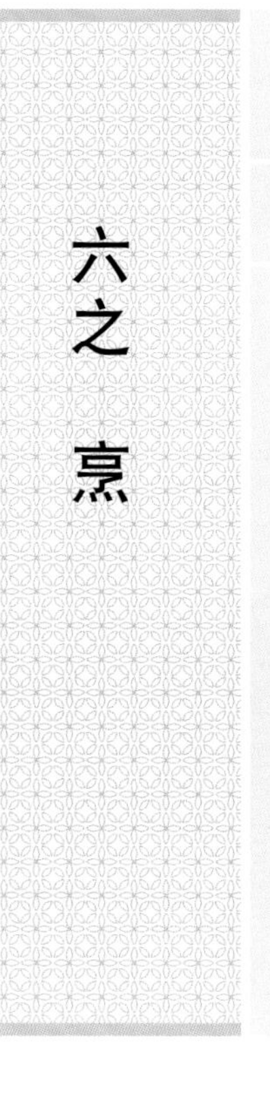

六之烹

Chapter 6 **BREWING**

烹茶之道,亦在阴阳相生也。阳茶阴泡,阴茶阳煮。阳气易散,略热即可释出;阴气内敛,非高温闷泡不可出。

Tea brewing should also follow the principle of balancing Yim and Yaŋ. The Yaŋ teas should be brewed in a Yim way, that is, in water with relatively lower temperature. On the contrary, the Yim teas must be brewed in a Yaŋ way, which is to be brewed in hot water or even to be cooked. It is because the Yaŋ Pneuma is radiative, and a slight heating will be enough for its release. The Yim Pneuma is absorptive, and therefore, to release the Yim Pneuma, high temperature and high pressure is definitely necessary.

绿茶太阳,宜玻璃杯温浸,七八十度水温足矣。高温则太阳气速散,徒失其鲜,而激寒伤小肠。

The green teas with Grand Yaŋ pneumas can be brewed in open glasses at the temperature of 70 – 80℃. Too high temperature will drive the Grand Yaŋ pneumas out rapidly, and lose the taste of the green teas. Moreover, high concentration of Grand Yaŋ pneuma brings cold activity (inaction) for body cells and hurts the duodenum. The best way of green tea brewing is to draw hot water in three to five phases into tea leaves. It is better to pour the hot water into a copper pot with a small hole in the pot bottom above the tea, and thus, the water will drip into the tea drop by drop with the temperature

Figure 15. The principle of tea brewing is the balance between Yīm and Yáŋ.

and molecular polarity of the water reduced, which will protect the sensitive hydroxybenzenes in the green teas. In the first time, we just make the leaves wet. After half a minute, we can draw the water slowly to around two centimeters, shake the glass for one minute, then draw the water slowly to fill the glass with several pauses to reduce the water temperature. This is the best way to make the green teas tasty and healthy. The chrysanthemum flavor or broad bean flavor can only be felt in this way.

青茶阳明,可以九十度水温于盖碗中闷香。盖碗通透,合于阳明之道。

The blue teas with Middle Yaŋ pneumas can be brewed in covered bowls at the temperature of around 90℃ to release the thick fragrance. The bowls are just covered loosely without extra pressure, which is good for Middle Yaŋ pneuma. One can also draw some hot water onto the cover to seal the soup, keep it covered until good fragrance flows out.

红茶少阳,则以九十七度以上水温速冲,热量速散,不使茶酰胺降解,薄胎黑陶壶最佳。低温则胺不可释,热久则茶褐素生而香失。若壶具施釉不透气,少阳气如少男受拘,力久困而酸,则汤味酸。

The red teas with Little Yaŋ pneumas need to be brewed at the temperature higher than 97℃ to dissolve the amides. But the amides will degrade if kept in high temperature for minutes. Therefore, we need to pour the boiled water into red tea, and cool it immediately.

The pot must be made of a material with good ventilation capacity, or the Little Yaŋ pneuma, just like a boy locked in a jail, will be destroyed. In this case, the soup will be sour. Thus, the best tea pot for red tea must be made of black pottery with high thermal conductivity and breathability.

黄茶少阴,可于厚壁白瓷壶中以九十五度水温冲饮。黄酮类得孔必入,故不可以陶壶紫砂之类多孔茶具,若使少阴气隐匿,如少女含羞,茶汤必涩。

The yellow teas with Little Yim pneumas, can be brewed in a porcelain pot with a thick wall. The temperature of the water must be higher than 95℃. Earthenware without glaze cannot be used for Yellow tea. Because there are a large number of tiny holes in the walls of earthenwares, which will lock the flavone of the Yellow tea. The Little Yim pneuma is like a girl who is too shy to see people. When it is locked, the tea soup will get astringent.

黑茶厥阴,紫砂壶一百度水温闷泡最佳。茶苷浓稠,溶于汤中;杂质细小,吸于壁孔内。故得汤香滑。若杂质过多,则可先润洗一遍。

The black teas with Middle Yim pneumas can be brewed in a red stone pot in 100℃ water. The thick concentration of glycoside will then be dissolved in the water, while the impurities such as theophylline will be locked by the holes of red stoneware. In this way, the tea soup will be smooth and sweet. If there are too many impurities, we can wash the tea with warm water before brewing.

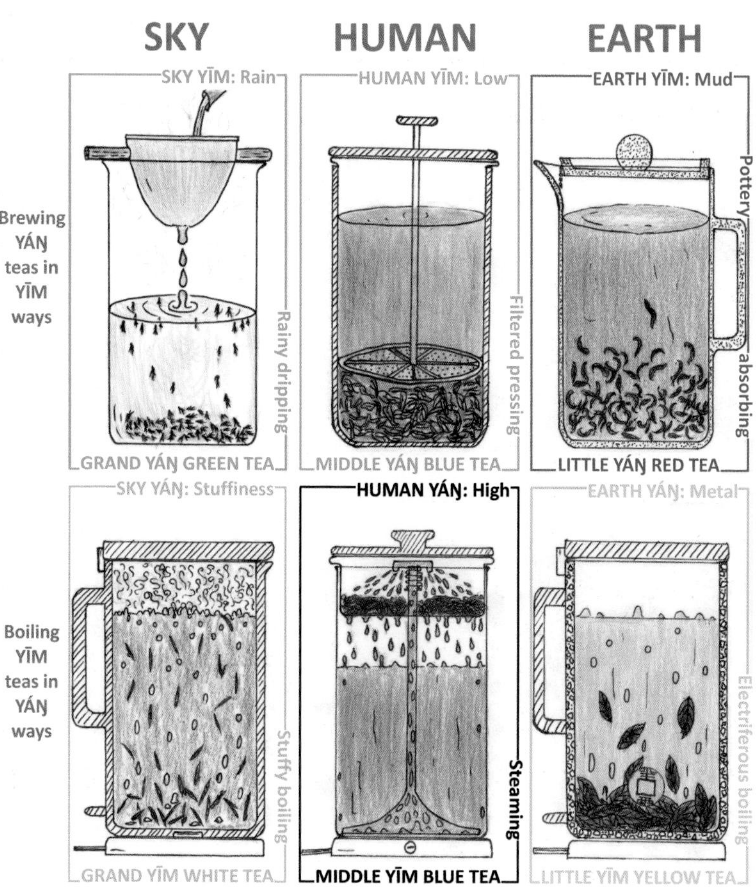

Figure 16. Simplified methods of tea brewing

白茶太阴,则必于煮水壶中直接煮沸方可成其汤气。更利于酯化反应,使太阴气愈足。

The white teas with Grand Yim pneumas, must be boiled to release the nutrition. During boiling, esterification reaction occurs between fulvic acid and theophylline, which makes the Grand Yim pneumas.

三阴之茶,皆可烹煮,其味醇厚。黑茶最宜蒸淋之法,可洗取多糖,得其精华,香甜滑腻。

The three Yim teas can all be boiled for drinking, which makes the soups much smoother and sweeter. The black tea can be leached by steam as the best way of brewing. The glycoside is leached off to the soup and is broken into shorter pieces. That is why the soup is sweeter.

七之 时

Chapter 7 SCHEDULE

饮茶之时,必合于人体之生物节律。其中昼夜轮替之节律最甚,人体经络脏器随昼夜而顺动,古者谓之子午流注。十二正经,各主一脏器而通手足,两两成对,而合为六正脉,其中行气阴阳各异,恰应于茶气之阴阳六类。经络顺昼夜节律而轮作,故饮茶之类亦必应之为宜。

The time schedule of tea drinking in one day must follow the biological clock of the human body. The rotation of day and night is the largest factor of the biological clock. The meridians and organs will work or be excited in turn during the rotation of day and night, which is called Midnight-Noon Ebb-Flow Doctrine. There are twelve major channels. Every channel controls an organ and leads either to hands or to feet. Every one hand channel or foot channel is linked with a full meridian. Thus, totally six meridians can be found, with either Yim or Yaŋ pneumas going inside. The six kinds of pneuma contained by the six kinds of tea go to their corresponding meridians. The meridian channels open in turn obeying the Doctrine every day. Therefore, the time schedule of tea drinking should also follow the Doctrine.

辰时方起,以绿茶太阳气入手太阳小肠经或足太阳膀胱经,提神利尿,释放一夜过滤沉淀之污垢,神清气足以作一日之计。

In the morning from seven to nine o'clock, we need to refresh up with the Grand Yaŋ pneumas of green teas. Pneumas of green teas go to Hand · Grand Yaŋ · Duodenum channel or Foot · Grand Yaŋ · Bladder channel, which is refreshing and diuretic. Metabolic wastes accumulated through the night can be washed out after drinking and one will be refreshed and energized for a new day.

巳时繁杂,以白茶太阴气入手太阴肺经或足太阴脾经,润肺健脾,外祛污浊之气,内强运化免疫之功,以候午时进食杂物。

During the busy time from nine to eleven o'clock, we need to strengthen the lungs to drive away dirty air, and to enhance the immune system, namely, the spleen, by drinking white tea with Grand Yim pneuma, which goes into Hand · Grand Yim · Lung channel or Foot · Grand Yim · Spleen channel.

午时进餐,以黄茶少阴气入手少阴心经或足少阴肾经,通心洗肾,排除进餐后升高之血糖,不使其化作血脂积于心血管。

At noon from eleven to thirteen o'clock, especially after lunch, blood sugar reaches the highest value of a day. The yellow tea is the best choice for this time, which carries the Little Yim pneuma going to Hand · Little Yim · Heart channel or Foot · Little Yim · Kidney channel, and degrade the blood sugar to clear the circulatory system, avoiding the blood sugar to be transformed into blood fat, which can block the vessels.

未时困倦,以红茶少阳气入手少阳三焦经或足少阳胆经,养颜利胆,调理内分泌以静心神,分泌胆汁以化积腻。

In the afternoon from thirteen to fifteen o'clock, the fat food eaten at noon has just gone through the stomach and reached the duodenum. The secretion of bile is then helpful to digest the fat, which can be enhanced by drinking red tea carrying Little Yaŋ pneuma, which go to Hand · Little Yaŋ · Three-gland channel or Foot · Little Yaŋ · Gallbladder channel. The three glands of pituitary, thyroid, and adrenal glands, will make people happy and optimistic when are enhanced. Drinking the red tea for three glands makes people comfortable at this most tiring time.

申时匆忙,以青茶阳明气入手阳明大肠经或足阳明胃经,清肠排毒,消解幽郁,化除内结,以利休憩。

The late afternoon from fifteen to seventeen o'clock is also a busy time. The food taken at noon has arrived at the colon, thus, the blue tea with Middle Yaŋ pneuma going to Hand · Middle Yaŋ · Colon channel or Foot · Middle Yaŋ · Stomach channel will be the best choice to clear the colon and get relaxed.

酉时壅塞,餐后以黄茶少阴气入足少阴肾经,可通肾清淤,化除陈垢。

From seventeen to nineteen o'clock after dinner is also a good time for yellow tea to wash the kidneys.

戌时悠闲,以黑茶厥阴气入手厥阴心包经或足厥阴肝经,安神疏肝,排解一日饮食积累于肝中之油脂,调理身心,轻松愉悦以利睡眠。

In the evening, drinking the black teas with Middle Yim pneuma going to Hand · Middle Yim · Thymus channel or Foot · Middle Yim · Liver channel is recommended. Those black teas for liver can clean the fat in liver and related organs such as prostate which makes it easeful for people to go sleeping. The black teas for thymus are even pertinent for good sleeping.

子午流注之中，阴经当令之时为该经收敛而盈，当补该经之阴气；阳经当令之时为该经耗散而虚，当补相应脏器之母，所谓“虚则补其母”。

The Midnight-Noon Ebb-Flow Doctrine is a schedule of active time of the twelve channels (Table 3). (This Doctrine has been accepted in China for thousands of years, and recently was confirmed by experiments in our laboratory.)

Every two hours, there will be a channel in active status. However, it is not simply to drink the tea that contains the pneuma going to the active channel. For the active Yim channel one can simply drink the tea going to this channel, for it absorbs pneuma. For the active Yaŋ channel, drinking the corresponding tea going to the active channel is helpless, because this channel is releasing pneuma. In Chinese Medicine, when an organ is losing pneuma and getting weak, one needs to nourish its "mother", that is another organ generating this active organ according to the Five Elements Theory (→Metal →Water →Wood →Fire →Soil→).

Table 3 Tea Drinking Schedule according to Midnight-Noon Ebb-Flow Doctrine

	Chinese Time	Time	Active Channel in Doctrine	Element	Generation / Restriction	Organ to nourish	Tea
Night	Zi	23:00 – 1:00	Foot · Little Yaŋ · Gallbladder	Wood	Metal	Lung	White
	Chou	1:00 – 3:00	Foot · Middle Yim · Liver	Wood	—	Liver	Black
	Yin	3:00 – 5:00	Hand · Grand Yim · Lung	Metal	—	Lung	White
Day	Mao	5:00 – 7:00	Hand · Middle Yaŋ · Colon	Metal	Soil	Stomach /Colon	Blue
	Chen	7:00 – 9:00	Foot · Middle Yaŋ · Stomach	Soil	Fire	Duodenum /Bladder	Green
	Si	9:00 – 11:00	Foot · Grand Yim · Spleen	Soil	—	Spleen /Lung	White

continued

	Chinese Time	Time	Active Channel in Doctrine	Element	Generation / Restriction	Organ to nourish	Tea
Day	Wu	11: 00 – 13: 00	Hand · Little Yim · Heart	Fire	—	Heart	Yellow
	Wei	13: 00 – 15: 00	Hand · Grand Yaŋ · Duodenum	Fire	Wood	Gallbladder /Three-gland	Red
	Shen	15: 00 – 17: 00	Foot · Grand Yaŋ · Bladder	Water	Metal	Colon /Stomach	Blue
Night	You	17: 00 – 19: 00	Foot · Little Yim · Kidney	Water	—	Kidney	Yellow
	Xu	19: 00 – 21: 00	Hand · Middle Yim · Thymus	Water	—	Thymus /Liver	Black
	Hai	21: 00 – 23: 00	Hand · Little Yaŋ · Three-Glands	Fire	Water	Kidney /Three-gland	Yellow/Red

卯时大肠经当令，大肠属金，土生金，胃属土，故宜胃经青茶。

For example, in the morning from five to seven o'clock, the colon channel is active. Colon is Yaŋ and Metal. Metal is generated by Soil. The organ supported by Yaŋ pneuma and Soil element is stomach. Therefore, the blue tea going to stomach is proper for drinking at this time.

辰时胃经当令，胃属土，火生土，小肠属火，故宜小肠经绿茶。

From seven to nine o'clock, the stomach channel is active. Stomach is supported by Yaŋ pneuma and Soil element. Soil is generated by Fire. The Yaŋ Organ of Fire is duodenum. Therefore, the green teas going to the duodenum will be a good choice.

未时小肠经当令，小肠属火，木生火，胆属木，故宜胆经之红茶。

In the afternoon from thirteen to fifteen o'clock, the duodenum of Fire is active, then we need to nourish the gallbladder of Wood to generate the Fire. That is why we prefer drinking red tea in the afternoon.

申时膀胱经当令，膀胱属水，金生水，大肠属金，故宜大肠经之青茶。

From fifteen to seventeen, the bladder of Water is active, then we need to nourish the colon of Metal by drinking blue tea.

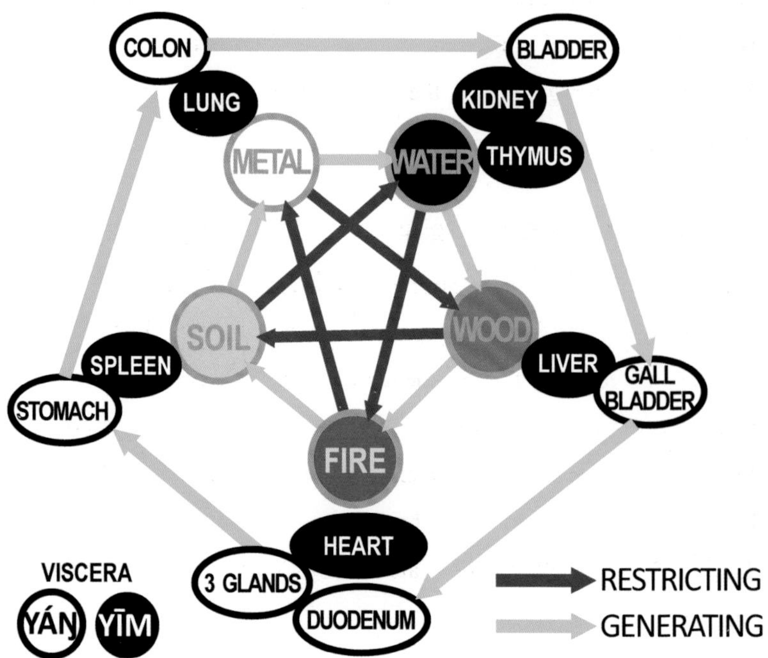

Figure 17. The Five Elements Theory suggests that organs generate each other in turn.

自酉至寅为夜,宜用阴茶。虽有少阳脉当令之时,亦可以阴茶用之,以收敛其相克之性而养之。

From seventeen to five o'clock of the next day, it is night defined by the Doctrine. We should mostly drink Yim teas. Even if it is the Yaŋ channel active time, we can choose the corresponding Yim channel going through the organ of certain element restricting the element of the active organ.

如亥时三焦经当令,三焦属火。如未眠,可饮肾经黄茶。肾经为阴属水,水克火,水收则火安。如尚劳,亦可用三焦经红茶。

For example, from twenty-one to twenty-three o'clock, it is the Three-gland time of Fire element. The best choice, if we are not going to sleep, is the Kidney Yellow tea with Yim pneuma and Water element. Thus, the water element will be absorbed into the Little Yim meridian and will not release to restrict the Fire in the Three-gland channel. One can also drink the Three-gland Red Tea if still busy and active.

虽六茶合于六时辰,然则常人一日不宜过三,过则恐伤神。最宜之时,亦非仅宜之时也,他时亦有宜饮者。愈阴者,宜时愈多,此"知白守黑"之理也。

Although the six types of tea have their best time for drinking, usually one should not drink more than three types of tea in the same day or it will hurt one's mind. Then what if one wants to drink tea all the day including the time beyond the chosen three teas. Actually, the best time is not the only time proper for drinking the corresponding

Figure 18. The extended tea selection schedule in a day according to Midnight-Noon Ebb-Flow Doctrine and the Five Elements Theory.

tea. It will definitely be fine to drink a tea beyond the best time but there is another rule. The Yim teas are proper for more times than the Yaŋ teas.

太阴白茶,日夜宜饮;厥阴黑茶,唯卯辰不宜,以免抑阳;少阴黄茶,自巳至亥时可饮,余时血糖本低,不宜多饮;少阳红茶,自辰至申时可饮;阳明青茶,寅卯可以清肠,未申可以舒胃;太阳绿茶,唯卯至辰可提神耳。

The Grand Yim white tea is proper for all day long. The second best is the Middle Yim black tea which can be drunk except from five to nine in the morning when one should not be sleepy. The Little Yim yellow tea is good after meals from nine to twenty-three o'clock. Before seventeen, the Heart yellow tea is good, and after seventeen, the Kidney yellow tea will be a better choice. The other time is not good for yellow tea because the blood sugar is not high. The Little Yaŋ red tea can be drunk from seven to seventeen o'clock. The gentle Little Yaŋ pneuma is always comfortable for the daytime. The Middle Yaŋ blue tea can be drunk either in the morning from three to seven o'clock to wash the colon, or in the afternoon from eleven to seventeen o'clock to comfort the stomach. The Grand Yaŋ green tea is only good for the morning generally to refresh the body and release the urine accumulated through the night.

八之季

Chapter 8 SEASONALITY

节律之次者四季轮替也。八之季中:“四季轮替也。”后面加上“《内经》谓春生夏长秋收冬藏,春季生肝,夏季生心,秋季生肺,冬季生肾。又曰:逆春气,少阳不生,肝气内变;逆夏气,太阳不长,心气内洞。此言四季养生当以表里相生。肝胆、心小肠、肺大肠、心包肾阳皆互为表里,故可以胆养肝,以小肠养心。”

The second factor of biological clock is the rotation of four seasons.

春生夏长秋收冬藏,以少阳红茶生发春日阳气,以太阳绿茶助长夏日精魄,以阳明青茶收敛秋日清神,以厥阴黑茶滋养冬日幽思。

The human body grows in different motives during different seasons. It wakes up in spring, grows quickly in summer, organizes in autumn, and rests in winter. Therefore, we need to drink more of a certain tea for the corresponding season. In spring, Little Yaŋ red tea can wake the Yaŋ pneuma of the body and the young spirits in the meantime. In summer, the Grand Yaŋ green tea can enhance the spirit and quicken the metabolism of the body. In autumn, the Middle Yaŋ blue tea can clear up the body and concentrate the spirit. In winter, the Middle Yim black tea can calm down the spirit and let the body have a rest. In the *Yellow Emperor's Canon of Medicine*, a principle of meridian loop was applied to the seasonal health protection. For example, the liver grows quickest in spring, and as the liver channel and the gallbladder channel are connected in a loop, nourishing the gallbladder channel will irrigate the root of the liver channel which is the most proper way to help the liver grow. Similarly, we could nourish the duodenum to irrigate the heart channel in summer, nourish the colon to irrigate the lung channel in autumn, and nourish the thymus to irrigate the three-gland channel (here, adrenal glands). That is why we should drink more red teas in spring, more green teas in summer, the blue teas in autumn, and the black teas in winter.

四季之间各有一时曰长夏,可以黄茶消解前季余疳。

Between every two seasons, there is a half month called Long-Summer, which means human body will get active and super sensitive. Thus, in four Long-Summers, the metabolic wastes are quickly accumulated, which need to be eliminated by yellow tea.

白茶提升免疫,则可四季常备也。

The white tea can significantly enhance the immune system, and therefore, is good for drinking in every season.

故曰:春红,夏绿,秋青,冬黑,长夏黄,四季白。

Therefore, there is a pithy formula for seasonality of tea choice: red tea for spring, green tea for summer, blue tea for autumn, black tea for winter, yellow tea for interseason, and white tea for all seasons.

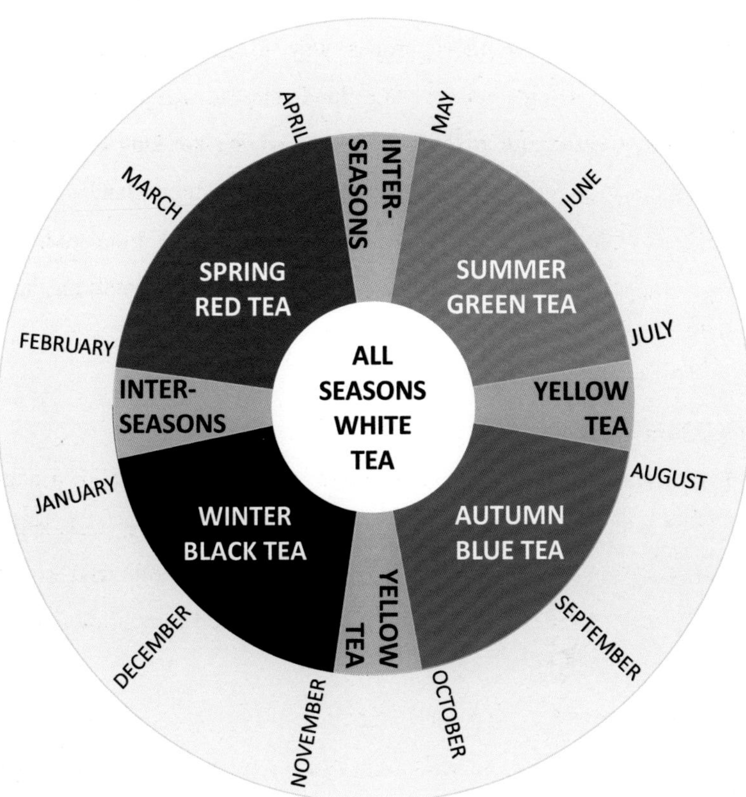

Figure 19. Teas suitable for different seasons.

To drink more of the right tea in the right season can make the body growth catch up with the season changing and avoid discomfort. In Chinese calendar, one year is divided into 24 solar terms. The days in between are called Node Pneumas (Jie Qi), which means the pneuma of the nature changes. And therefore, the best choice of tea changes accordingly (Table 4). For example, the Great Cold Node pneuma in late January is the coldest day of the year, and the immunity of human body touches the lowest point. Therefore, the Spleen White Tea, Shoumei Tea (Old Eyebrows), will be the best choice for Great Cold to enhance immunity. The Great Heat Node Pneuma in late July is the hottest day of the year, and very stuffy in East Asia. Thus, the Lung White Tea, White Silver Needle Tea, will be the best choice for Great Heat to ventilate the lungs.

表 4　二十四节气选茶指南

节气	日期	茶　种	茶类	经　络	产　　地
立春	二月上	雒越红	红	手少阳三焦经	广西-凌云-岑王老山
雨水	二月下	正山小种	红	手少阳三焦经	福建-崇安-武夷山
惊蛰	三月上	苗红	红	足少阳胆经	贵州-绥阳-大娄山
春分	三月下	葡红	红	足少阳胆经	云南-永平-博南山
清明	四月上	君山金砖	黄	足少阴肾经	湖南-岳阳-君山
谷雨	四月下	白牡丹	白	足太阴脾经	福建-福鼎-太姥山
立夏	五月上	碧螺春	绿	手太阳小肠经	江苏-苏州-洞庭山
小满	五月下	龙井	绿	足太阳膀胱经	浙江-杭州-狮峰

续　表

节气	日期	茶　种	茶类	经　络	产　　地
芒种	六月上	黔绿珠	绿	足太阳膀胱经	贵州-湄潭-大娄山
夏至	六月下	猴魁	绿	手太阳小肠经	安徽-太平-黄山
小暑	七月上	金锭茶	黄	足少阴肾经	云南-芒市-老中山
大暑	七月下	白毫银针	白	手太阴肺经	福建-福鼎-太姥山
立秋	八月上	铁观音	青	手阳明大肠经	福建-安溪-枫凤山
处暑	八月下	凤凰单枞	青	足阳明胃经	广东-潮安-凤凰山
白露	九月上	大红袍	青	足阳明胃经	福建-崇安-武夷山
秋分	九月下	齐鲁青未	青	足阳明胃经	山东-莱芜-龙山
寒露	十月上	德昂酸茶	黄	足少阴肾经	贵州-正安-大娄山
霜降	十月下	雪梨银针	白	手太阴肺经	云南-景谷-无量山
立冬	十一月上	明月夜黑铁	黑	足厥阴肝经	福建-安溪-枫凤山
小雪	十一月下	六堡茶	黑	手厥阴心包经	广西-苍梧-大桂山
大雪	十二月上	金花黑茶	黑	足厥阴肝经	湖南-永州-阳明山
冬至	十二月下	普洱熟茶	黑	手厥阴心包经	云南-澜沧-景迈山
小寒	一月上	梵金髻	黄	手少阴心经	贵州-江口-梵净山
大寒	一月下	寿眉	白	足太阴脾经	福建-福鼎-太姥山

Table 4 Teas for 24 Solar Terms

Solar Term	Dates	Tea	Type	Channel	Locality
Establishing Spring	Early February	Luoyue Red Tea	Red	Hand · Little Yaŋ · Three-gland	Cenwanglao Mountain, Lingyun, Guangxi
Rain Drops	Late February	Lapsang-Souchong	Red	Hand · Little Yaŋ · Three-gland	Wuyi (Bohea) Mountain, Chongan, Fujian
Hibernation Awakening	Early March	Meitan Red Tea	Red	Foot · Little Yaŋ · Gallbladder	Dalou Mountain, Suiyang, Guizhou
Spring Equinox	Late March	Red Grape	Red	Foot · Little Yaŋ · Gallbladder	Bonan Mountain, Yongping, Yunnan
Clear Brightness	Early April	Junshan Golden Brick	Yellow	Foot · Little Yim · Kidney	Junshan Hill, Yueyang, Hunan
Grain Rain	Late April	White Peony	White	Foot · Grand Yim · Spleen	Taimu Mountain, Fuding, Fujian
Establishing Summer	Early May	Green Snails	Green	Hand · Grand Yaŋ · Duodenum	Dongting Hills, Suzhou, Jiangsu
Grain Full	Late May	Dragon Well	Green	Foot · Grand Yaŋ · Bladder	Lion Peak, Hangzhou, Zhejiang

continued

Solar Term	Dates	Tea	Type	Channel	Locality
Awn on Seed	Early June	Guizhou Turquoise Pearls	Green	Foot · Grand Yaŋ · Bladder	Dalou Mountain, Meitan, Guizhou
Summer Solstice	Late June	Monkey Champion	Green	Hand · Grand Yaŋ · Duodenum	Yellow Mountain, Taiping, Anhui
Slight Heat	Early July	Bud-Chem Golden Brick	Yellow	Foot · Little Yim · Kidney	Fanjing Mountain, Jiangkou, Guizhou
Great Heat	Late July	White Silver Needles	White	Hand · Grand Yim · Lung	Taimu Mountain, Fuding, Fujian
Establishing Autumn	Early August	Titkuanyim	Blue	Hand · Middle Yaŋ · colon	Fengfeng Hills, Anxi, Fujian
Limit of Heat	Late August	Phoenix Unique	Blue	Foot · Middle Yaŋ · Stomach	Fenghuang Mountain, Chao'an, Guangdong
White Dew	Early September	Red Garment	Blue	Foot · Middle Yaŋ · Stomach	Wuyi (Bohea) Mountain, Chongan, Fujian
Autumn Equinox	Late September	Cheerio Ching	Blue	Foot · Middle Yaŋ · Stomach	Dragon Mountain, Laiwu, Shandong

continued

Solar Term	Dates	Tea	Type	Channel	Locality
Cold Dew	Early October	De'ang Lactics	Yellow	Foot · Little Yim · Kidney	Laozhong Mountain, Mangshi, Yunnan
Frost Descent	Late October	Pear Silver Needles	White	Hand · Grand Yim · Lung	Wuliang Mountain, Jinggu, Yunnan
Establishing Winter	Early November	Dark Moon	Black	Foot · Middle Yim · Liver	Fengfeng Hills, Anxi, Fujian
Light Snow	Late November	Liubao	Black	Hand · Middle Yim · Thymus	Dagui Mountain, Cangwu, Guangxi
Great Snow	Early December	Golden Fungi	Black	Foot · Middle Yim · Liver	Yangming Mountain, Yongzhou, Hunan
Winter Solstice	Late December	Pre-fermented Pu'erh	Black	Hand · Middle Yim · Thymus	Jingmai Mountain, Lancang, Yunnan
Slight Cold	Early January	Golden Buns	Yellow	Hand · Little Yim · Heart	Fanjing Mountain, Jiangkou, Guizhou
Great Cold	Late January	Old Eyebrows	White	Foot · Grand Yim · Spleen	Taimu Mountain, Fuding, Fujian

九之品

Chapter 9 MERIDIANS

成茶品者,其因素有三,曰工艺、曰原料、曰环境。工艺为上,因工艺之异,而有阴阳之别,成六类之茶。环境为下,好山好水,方可成好茶,然则不能异其品类之别。原料为中,以嫩芽肉叶所制者成气轻,上行而入手经;以多筋老叶所制者成气重,下沉而入足经。故六类茶因用料老嫩,成气轻重,又分十二品。

The qualities of the teas are determined mainly by three factors, i. e., fermentation technique, original material, and plantation environment. Definitely, fermentation technique is the most important factor. Different fermentations make different pneumas of Yim or Yaŋ, and produce six types of tea. Among the three factors, environment is the least important. A proper mountain with good earth and good water is the prerequisite to produce good tea, but does not change the type of tea. The original material, including the cultivation race of the teas, the picking time, and the part of the tree, is the second important factor. Genomes of different tea cultivars will make various fragrances, which means those teas contain different materials and may have different functions. There are two subspecies of China tea (*Camellia sinensis*). The Eastern subspecies is also known as Chinese tea (*Camellia sinensis* var. *sinensis*), and the Western subspecies is also known as Assamese tea (*Camellia sinensis* var. *assamica*). Two subspecies diverged more than 300 thousand years ago. Inside the subspecies, all the eastern teas came from an ancestor of more than 6,000 years ago, which has been found in the archaeological site Tianluoshan of Zhejiang. All the Western teas came from an ancestor tree of around 3,000 years ago in Yunnan. There have been a large number of interbreedings, including the Cambodian tea and the red tea cultivars in Yunnan and Guizhou. On another side, the material in the bud of early spring will differ far from the late leaves, and even farther from those of autumn. Therefore, some teas made by early buds release light pneumas going to the hand channels, while others made by late leaves might release heavy pneumas going to the foot channels. In this case, the six types of tea are further divided into twelve kinds.

太阳绿茶者,碧螺、猴魁芽薄气轻,上升于小肠经而提神;乌牛、龙井芽厚气重,下沉于膀胱经而利尿。

Among the Grand Yaŋ green tea, Pillutsen (Green Snails), Houkui (Monkey Champion), etc., made of thin buds growing at relatively low temperature, will release light pneumas going to the Duodenum channel, and are good for refreshing. Longjing (Dragon Well), Wuniuzao (Black Buffalo), etc., made of thick buds growing at relatively high temperature, will release heavy pneumas going to the Bladder channel, and are good for releasing urine.

阳明青茶者,铁观音叶嫩气轻,上升于大肠经而清肠;岩茶叶老气重,下沉于胃经而解郁。

Among the Middle Yaŋ blue teas, Titkuanyim and Frozen Peak Oolong, made of thin leaves oxidized at a low temperature, will release light pncumas going to the colon channel, and are good for cleaning it. Rocky teas like Dahongpao (Red Garment) and Fongfang-Dancong (Phoenix Unique), made of thick leaves oxidized at high temperature will release heavy pneumas going to the Stomach channel, and are good for relaxing the stomach.

少阳红茶者,雒越、正山芽细气轻,上升于三焦经而养颜;滇红、湄潭芽粗气重,下沉于胆经而化痞。

Among the Little Yaŋ red teas, Luoyue, Lapsang-Souchong, etc., made of thin buds of Eastern subspecies, will release light pneumas going to the Three-gland channel, and make people feel happy and cheerful. Red teas produced in Yunnan or Guizhou made of thick buds of Western subspecies will release heavy pneumas going to the Gallbladder channel and are good for digesting the fat.

少阴黄茶者,梵金髻叶嫩气轻,上升于心经而活血;君山、娄山叶老气重,下沉于肾经而洗肾。

Among the Little Yim yellow teas, Fanjinji (Golden Buns), made of thin leaves of Eastern subspecies, will release light pneumas going to the Heart channel, and is good for cleaning the blood. The Golden Bricks of Junshan or Loushan, made of thick leaves of Western subspecies, will release heavy pneumas going to the Kidney channel, and are good for cleaning the kidneys and medullae.

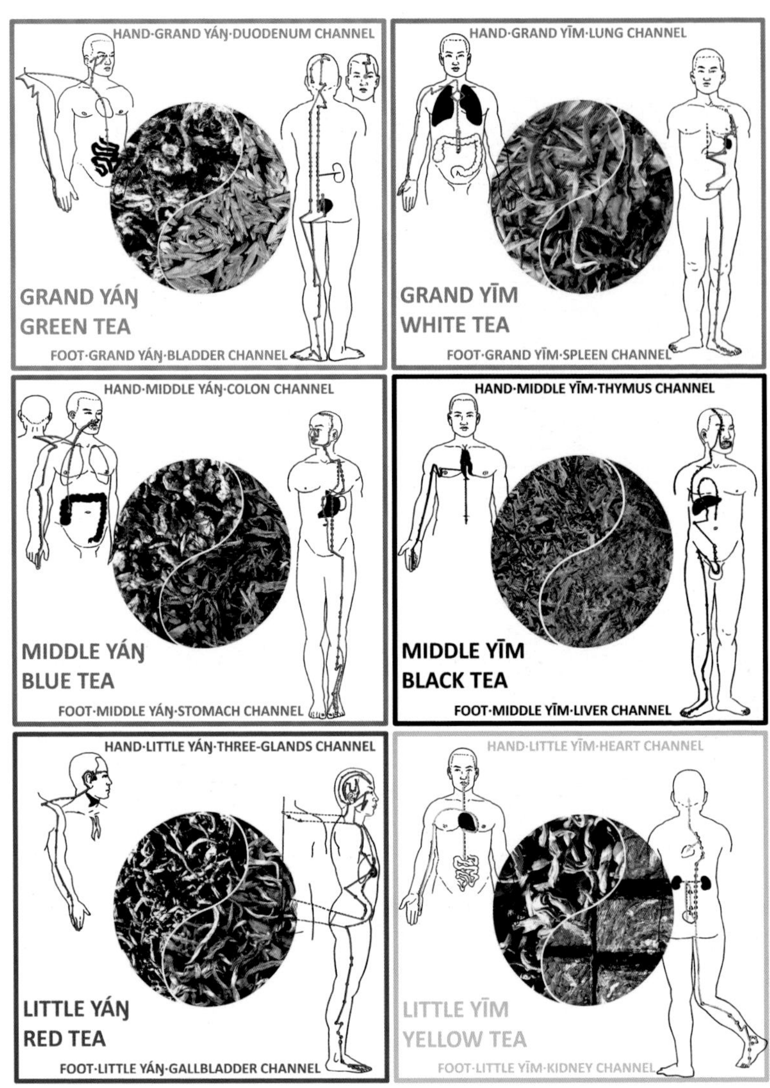

Figure 20. The meridian channels that the twelve kinds of tea go into.

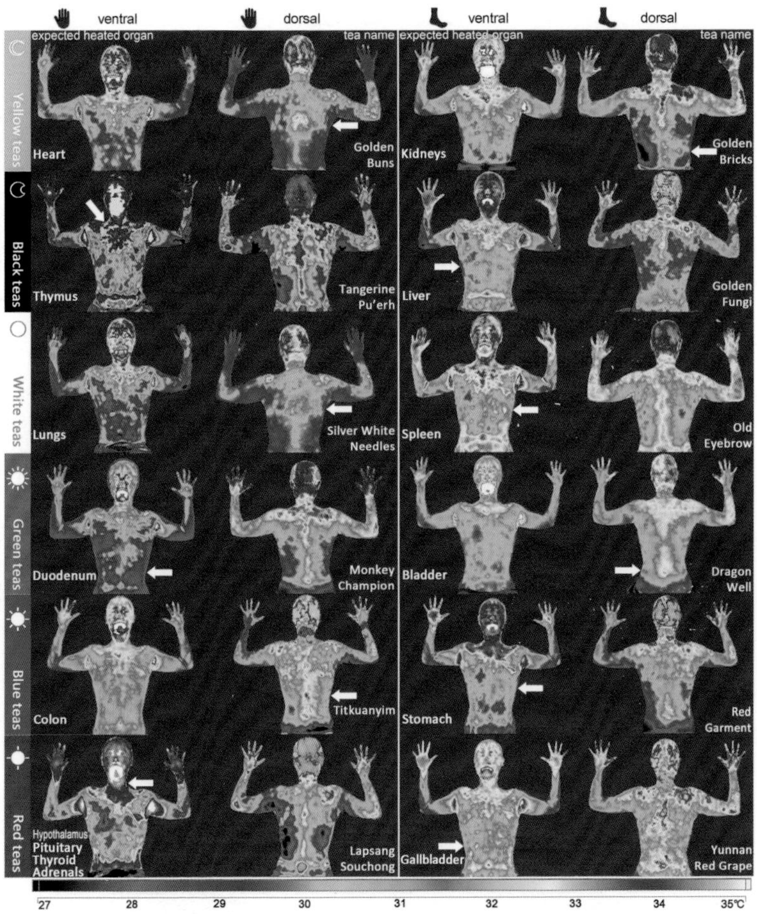

Figure 21. Infrared imageries after drinking 12 kinds of tea draining into 12 meridians

厥阴黑茶者,熟普、六堡叶嫩气轻,上升于心包经而安神;生普、金花叶老气重,下沉于肝经而解腻。

Among the Middle Yim black teas, Pre-fermented Pu'erh, Liubao, etc., made of thin leaves at a high temperature, will release light pneumas going to the Thymus channel, and are good for enhancing sleep. Post-fermented Pu'erh, Golden Fungi, etc., made of thick leaves at a low temperature, will release heavy pneumas going to the Liver channel, and are good for cleaning the fat accumulated in liver and other parts of the body.

太阴白茶者,白毫银针芽嫩气轻,上升于肺经而润肺;寿眉、叶老气重,下沉于脾经而健脾。

Among the Grand Yim white teas, White Silver Needle, made of early buds, will release light pneumas going to the Lung channel, and is good for cleaning the lungs. Shoumei (Old Eyebrows) and White Peony, made of old leaves, will release heavy pneumas going to the Spleen channel, and is good for enhancing the immune system.

表 5 各种茶的归经以及对应保健功效

茶类	茶 种	归 经	辅助治疗症状
绿茶	猴魁	手太阳小肠经	思维迟钝、积食不化。
绿茶	龙井	足太阳膀胱经	小便不畅、嘌呤过高、痛风、暑热痧症、中暑晕厥。
青茶	铁观音	手阳明大肠经	便秘、腹泻、肠道紊乱、食欲怪异、鼻窦炎。

续 表

茶类	茶 种	归 经	辅助治疗症状
青茶	大红袍、单枞	足阳明胃经	胃炎、胃溃疡、胃酸过多、烧心、消化不良、幽门感染。
红茶	正山小种	手少阳三焦经	情绪急躁、内分泌紊乱、甲状腺结节、甲亢甲减、脸侧过敏。
红茶	博南滇红	足少阳胆经	油腻不化、胆囊炎、胆结石、风邪入体、偏头痛、体侧酸痛、风寒感冒。
黄茶	梵金髻	手少阴心经	各类心血管疾病、心律不齐、心肌缺血、四肢冰凉、高血糖、糖尿病、高血脂、低血小板。
黄茶	金锭	足少阴肾经	肾结石、肾积水、腰酸背痛、足底冰凉、痛风、骨骼疼痛、骨质疏松、大脑蜕变、癫痫、高血压、冠状病毒感染。
黑茶	陈香熟普	手厥阴心包经	思虑过度、失眠、神经疼痛、神经衰弱。
黑茶	金花黑茶	足厥阴肝经	脂肪肝、前列腺炎、皮肤色斑沉积、眼疲劳、青光眼、高血脂。
白茶	白毫银针	手太阴肺经	咳嗽、哮喘、支气管炎、咽喉炎、肺气肿。
白茶	寿眉	足太阴脾经	免疫力低下、流感、疱疹、红斑狼疮、白癫风、皮炎皮疹、乳腺炎。

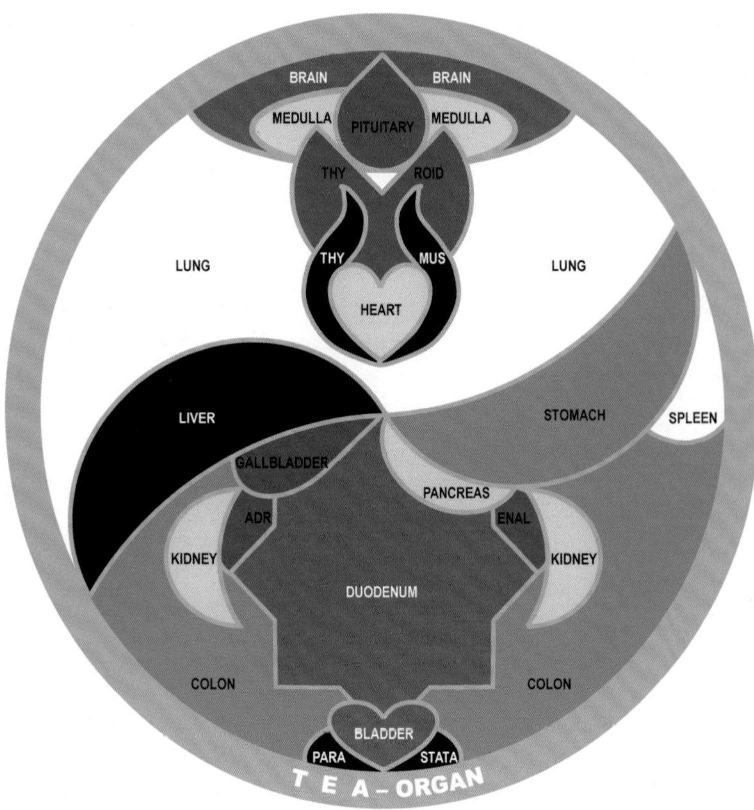

Figure 22. The organs that the six types of tea nourish.

Table 5 Meridian going of the representative teas and their assistant medical functions

Type	Tea name	Meridian	curable diseases
Green	Monkey Champion	Hand · Grand Yaŋ · Duodenum	slow thinking, indigestion
Green	Dragon Well	Foot · Grand Yaŋ · Bladder	unsmooth urination, high purin, heatstroke, gout
Blue	Titkuanyim	Hand · Middle Yaŋ · Colon	constipation, diarrhea, intestinal flora disorder, weird appetite, nasosinusitis
Blue	Red Garment, Phoenix Unique	Foot · Middle Yaŋ · Stomach	gastritis, gastric ulcer, acid indigestion, heartburn, indigestion, *Helicobacter pylori* infection
Red	Lapsang-Souchong	Hand · Little Yaŋ · Three-gland	anxiety, endocrine disturbance, thyroid nodule, hypothyroidism
Red	Bonam Red Grape	Foot · Little Yaŋ · Gallbladder	fat indigestion, cholecystitis, gall-stone, wind-evil, cephalagra, body-side pain, common cold
Yellow	Golden Buns	Hand · Little Yim · Heart	angiocardiopathy, arrhythmia, myocardial ischemia, cold hands and feet, hyperglycaemia, diabetes, hyperlipemia, thrombocytopenia

continued

Type	Tea name	Meridian	curable diseases
Yellow	Golden Brick	Foot · Little Yim · Kidney	kidney stone, uronephrosis, backache, gout, cold feet, ostalgia, osteoporosis, encephalanalosis, epilepsia, hypertension, coronavirus infection
Black	Pre-Fermented Pu'erh	Hand · Middle Yim · Thymus	mania, agrypnia, neuralgia, panasthenia
Black	Golden Fungi	Foot · Middle Yim · Liver	hepatic adipose infiltration, prostatitis, skin stain deposit, kopiopia, glaucoma, hyperlipemia
White	White Silver Needles	Hand · Grand Yim · Lung	cough, asthma, bronchitis, iaryngopharyngitis, emphysema
White	Old Eyebrows	Foot · Grand Yim · Spleen	hypoimmunity, influenza, herpes, lupus erythematodes, vitiligo, eczema, itch, mammitis

天之气,阳者升,阴者降。人之气,阴者升,阳者降。人气逆天气而动,方可成循环,得天气之利,而养人气之本。诸茶最能归经走气,畅通诸脉,故为养人之宝也。

In the nature, the Yaŋ pneumas go upwards and the Yim pneumas go down. But in human bodies, the Yaŋ pneumas go down and the Yim pneumas go up. Therefore, the pneumas go in an opposite direction inside and outside the body, which makes a full

Curative effect of heart yellow tea (Golden Buns) on a late stage sample of diabetes in six months

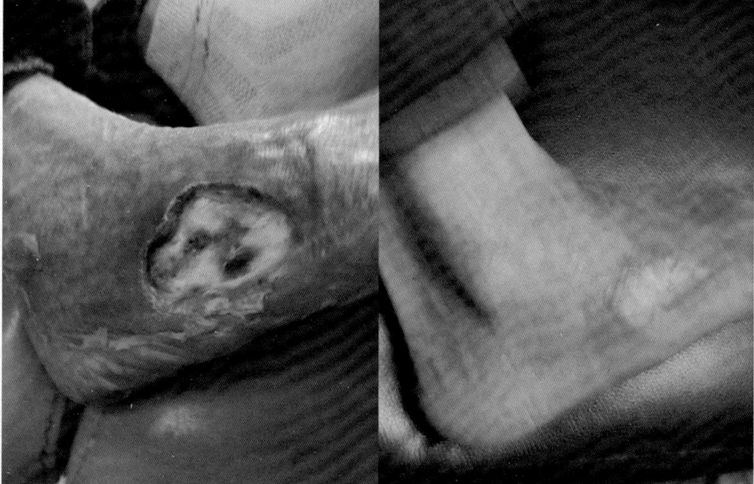

Curative effect of kidney yellow tea (Golden Bricks) in the evening and bladder green tea (Green Pearls) in the morning on a sample of gout in two days.

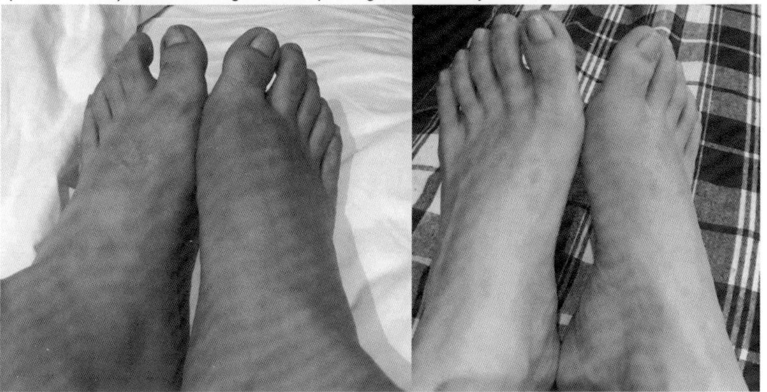

Figure 23. Two samples of the curative effects of proper teas.

circle, and drives the pneumas of the Nature into human body to nourish our souls. The teas release most of the pneumas for the human. Drinking tea can well open the meridians, and enhance the health. However, only when one drinks the teas in temperature higher than that of human body (37℃), can the pneumas go into the meridians. The tea soup temperature of 54℃ is the best for pneuma activity.

十之辨

Chapter 10 RATING

茶气之善，于茶乃根本也！好茶必有好气。气之善者在三，一曰纯，二曰正，三曰强。

The quality of pneuma is most important for the teas. A good tea must contain good pneuma. To rate the quality of pneumas, three parameters are needed, i.e., purity, healthiness, and concentration.

纯气之茶，其气唯一，利养体魄。道家谓之得一，佛家谓之不二。气若有二，入体则行于异脉，经脉相交之处则是二气相冲之所，体魄必伤于此。以气之纯度分，则茶可有四等。

A pure tea contains only one kind of pneuma, and can nourish the body souls well. It is called "Acquiring One" in Taoism, or "No Two" in Buddhism. Our body cells are bathed in tissue fluids. The tissue fluids flow along the meridians in turn. A kind of pneuma can drive the flow of a certain meridian. If a tea contains more than one kind of pneuma, different flows will be driven after drinking. And when the flows cross, the body will be hurt at the very point. According to the purity of pneuma, tea can be rated into four grades.

一等之茶，其气入单经，是为极品。

First Grade, the pneuma goes into only one channel of one meridian. That is the best tea.

二等之茶，其气虽同属一脉，然则在手足二经之间游走。譬如绿茶，若气自手太阳小肠经过睛明穴，入足太阳膀胱经，则虽为太阳气，已分二气矣，不及其一气之善。同脉二经之余气入支路，亦或有相交相冲之处。

Second Grade, the pneuma goes into two channels of the same meridian. For example, some green teas make pneumas go from Hand · Grand Yaŋ · Duodenum channel, through Jing Ming point between eyes, into Foot · Grand Yaŋ · Bladder channel. Although the pneuma is Grand Yaŋ only, as it is divided into two channels, the effect is then weakened. This grade of tea is also a good tea, but some pneumas might go into branches of the channels and collide each other.

三等之茶，气分入二脉，而阴阳之性同。如今之花香红茶，本具少阳之气，兼有青茶阳明之气，同为阳气，而入异脉。异脉主路相交，则相冲之害甚矣。

Third Grade, the pneumas go into two different meridians of the same direction. For example, there are some kinds of red tea with high concentration of flower flavor. The red tea contains Little Yaŋ pneuma originally, and thus, is added with Middle Yaŋ pneuma of flower flavor. Although they are both Yaŋ pneumas going in the same direction in the body, the meridians crisscross around the neck and the cardia where the body might be hurt.

四等之茶，其气兼有阴阳，入阴阳异脉，其气相向而冲，则害之极。如普洱生茶之新成者，其气以青茶之阳明为多，而兼有黑茶之

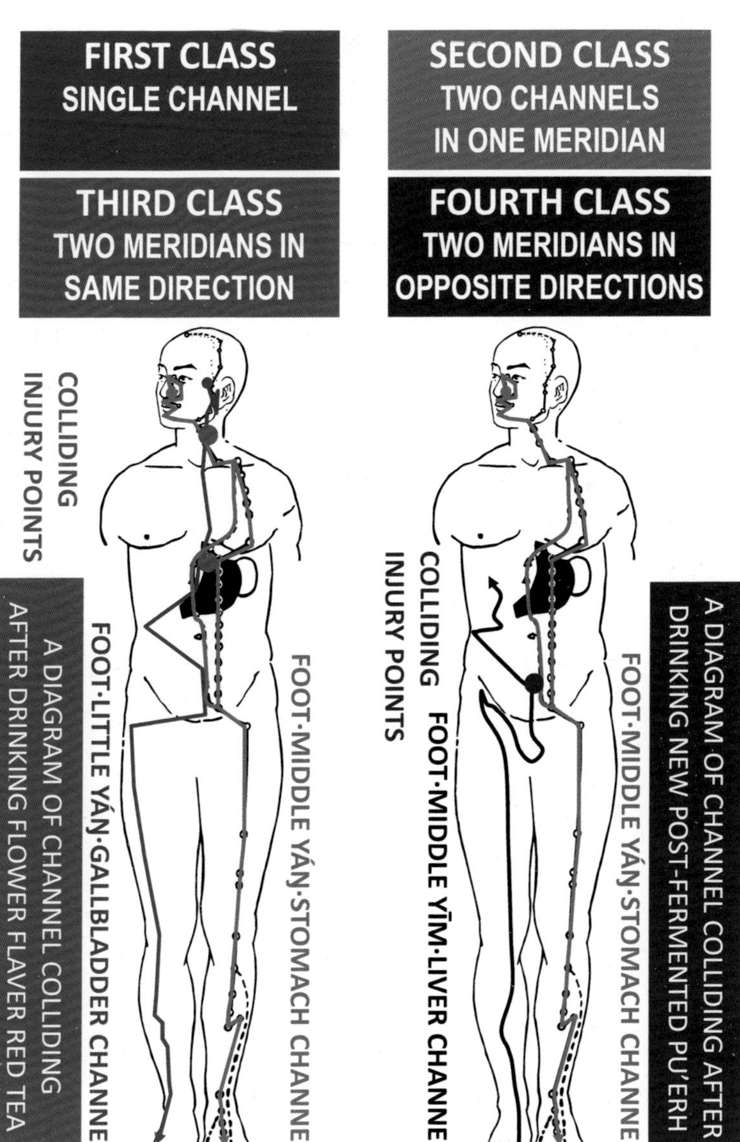

Figure 24. Rating principle of the teas and the mechanism of channel colliding.

厥阴,阳明气走胃经经腹部下足,厥阴气走肝经自足经腹部上目。此时若饮之,二气相冲于腹,害重者腹痛如绞。故生普必须久置经年,待阳明气发散尽,厥阴气积蓄满,则成一等茶矣。

Fourth Grade, the pneumas go into two channels of opposite directions. The tissue fluids then conflict heavily and the body will be hurt severely. For example, the new production of Post-Fermented Pu'erh, which is a kind of semi-manufacture, contains both Middle Yaŋ pneuma of blue tea and Middle Yim pneuma of black tea. The Middle Yaŋ pneuma goes down into the Stomach channel, through the hypogastrium to the legs. The Middle Yim pneuma goes up into the Liver channel, through the hypogastrium to the eyes. Two pneumas with the tissue fluids are then collided at the hypogastrium after drinking this tea, abdominal pain rises in severe cases. Therefore, the Post-Fermented Pu'erh must be kept for years before drinking to release the Yaŋ pneuma and to absorb the Yim pneuma to the highest concentration, which makes the Pu'erh a First Grade Tea.

正气之茶,饮之有补益,增补经络之气而强体魄。若气有邪,饮之则破气,经络气亏则体魄伤矣。

A healthy tea contains the pneuma good for the human body. Good pneumas can be added to the meridians and enhance the souls. If the pneuma is bad, the meridians will be destroyed and the body souls will be hurt after drinking.

食药之类,气皆有正邪之分。如韭蒜之类气邪,食之使血气凝滞,更破少阴之本;霉腐之类气邪,食之使肝气郁结,而破厥阴之

Figure 25. Six evil pneumas to be avoided during tea making.

本。蕨蕺之类气邪,食之使脾气虚弱,遂破太阴之本。此寒暑燥三阴邪气也。

The pneumas contained in foods or medicines can all be good or bad. The bad pneumas are also classified into Yim and Yaŋ. For example, those from chives (*Allium fistulosum*, *A. tuberosum*) and garlics (*Allium sativum*) are Little Yim pneumas and will reduce the efficiency of blood oxygen stream. Those from mould are Middle Yim pneumas, and will hurt the liver. Those from fiddle heads (*Pteridium aquilinum*) and houttuynia (*Houttuynia cordata*) are Grand Yim pneumas and will hurt the spleen and the immune system. These are called three evil Yim pneumas, namely, Little Yim cold evil, Middle Yim sultry evil, and Grand Yim dry evil, respectively.

茶为天生嘉木,其气固正,然则种之不当,制之不慎,存之不妥,则气易邪。守正何难,入邪何易。“采不时,造不精,杂以卉莽,饮之成疾。”

Tea is a kind of nice tree naturally. Originally, it contains healthy pneumas. However, the pneumas will easily turn bad if the tea is not well planted, not well fermented, or well kept. Good keeping is extremely hard, while turning bad is terribly easy. “Picking at a wrong time, fermenting in a rough and slipshod way, or mixing with various weeds, will make the tea drinking morbigenous.”

种茶之地,“野者上,园者次”,何也?野者得山川草木天然之卫气,以转营气,一何正也。园者近人,人有私欲而不尽合天道,其气固杂,以渐茶气。若近道路工厂,或施农药,少阴寒邪沾污,更不可饮。

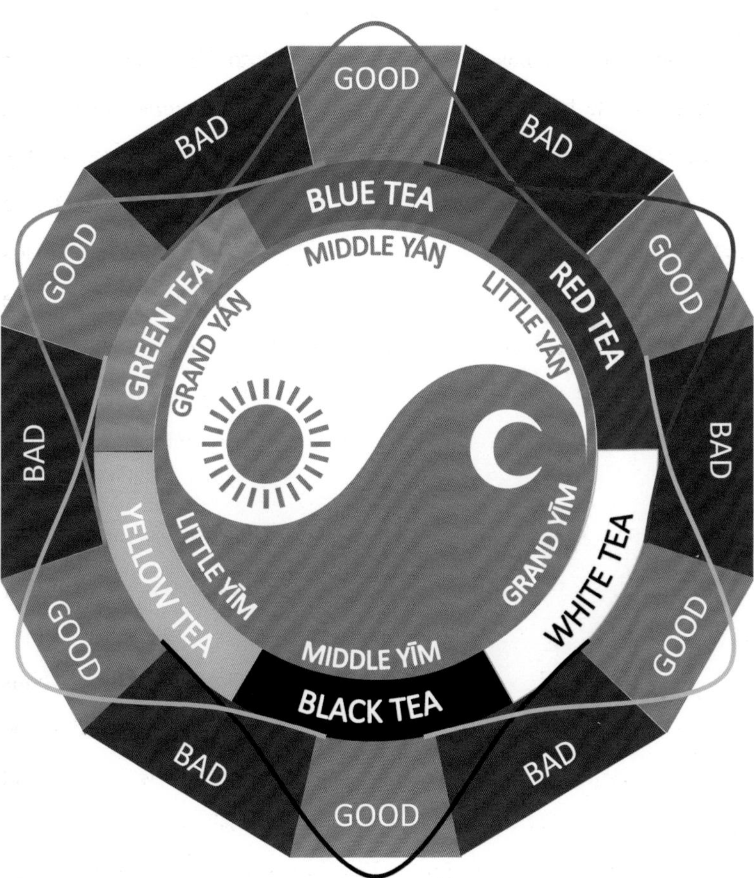

Figure 26. It is very important to seize the opportunities of the reaction peaks to make good teas.

For the place of tea plantation, "wild teas are always better than those from the gardens." That is because wild teas contain the nutritional pneumas transformed from protectional pneumas of the nature in mountains and forests. The natural pneuma must be the best. However, those tea trees in the gardens are too close to people. People always show selfishness which is not natural. Therefore, the tea pneumas are stained by the various human pneumas. If the gardens are close to the road or factory or pesticide, the teas are then stained by the Little Yim cold evil and inadequate fir drinking.

采茶之时,日曝雨淋,青料霉腐,更不当矣!须避厥阴暑邪之气。

When picking tea leaves, the tea may mildew and rot after exposed to the sun or the rain for too long. That is even worse as the Middle Yim sultry evil rises.

制茶为气正之关键,非室、器、水、风、工诸般皆净不可得正。卉莽之杂,太阴燥邪之侵也。

Fermentation is most essential to the healthy pneumas. The room, tools, water, air, and workers should all be clean enough. If the tea gets dirty or is mixed with weeds, the pneuma will turn into Grand Yim dry evil.

正茶出于净山净工,若存之不妥,则须臾废矣,潮湿污秽之所,断不可存茶。若阳茶腐,阴茶霉,断不可冲饮。阴茶易收外气,若收三阴邪气,虽丝毫亦废矣。幸者三阴邪气浊而下沉,秽而隐匿,故易避之。白茶日晒,黑茶高搁,黄茶锡封,则妥矣。

Healthy teas come from clean farms and clean factories. However, if not kept properly, they will turn bad very soon. The teas should never be kept in humid or dirty places. When the Yaŋ teas rot or the Yim teas mildew, they are not drinkable. The Yim teas always absorb the pneumas from outside. If they absorb any of the three Yim evil pneumas, they will be destroyed. Luckily, the three Yim evil pneumas are heavy, descending, dirty and hiding, which are easy to keep away. It will be fine enough to keep the tins of white teas in sunshine, the tins of black teas in the highest places, the tins of yellow teas sealed with metal.

强气之茶,毫滴未漏,可尽茶之利也。采之季,于清明前,芽叶未展,蓄气未耗,以成茶气必强。若明后,则愈晚气愈弱。采之时,须避风湿热之三阳邪气。狂风不可采以伤少阳,淫雨不可采以伤阳明,烈日不可采以伤太阳。若采之,气散半矣。

Thick teas contain high concentration of pneumas, without missing a slightest amount of them, and can make a full use of the materials. If we pick the teas before Qingming festival (~ April 5th), most of the buds are still closed, the pneumas accumulated during the winter have not been used, and therefore, we can get the thickest teas. After Qingming festival, the pneumas get thinner. When we pick the teas, the three Yaŋ evil pneumas must be avoided. The Little Yaŋ red teas must avoid the strong wind. The Middle Yaŋ blue teas must avoid the cold rain. The Grand Yaŋ green teas must evade the burning sun. That is to evade the Little Yaŋ windy evil, the

Middle Yaŋ wet evil, and the Grand Yaŋ hot evil. If the teas are picked in those weather conditions, most of the pneumas will dissipate.

制茶之法,乃截阴阳转换之机也。不可失机,急缓均不可得。绿茶采之即须杀酶以封太阳营气,不使稍转。太阳以化少阴,则得黄茶。少阴之机,稍纵即逝,故黄茶之法固难,不及则绿茶太阳气尚存,过之则黑茶厥阴气遽生,气乱则散,不足饮也。

Tea making needs to seize the right moments of transformation between Yim and Yaŋ, neither too early nor too late. For the green tea, the enzymes must be inactivated right after picking, to keep the Grand Yaŋ pneuma, and not any transformation. The Grand Yaŋ pneuma can be transformed into Little Yim pneuma, and the yellow tea is thus made. The moment of Little Yim emergence is shortest, and therefore, the yellow tea is most difficult to make. Before the moment, the Grand Yaŋ pneuma of green tea remains. After the moment, the Middle Yim pneuma of black tea emerges. Admixture of different pneumas is bad for health and unacceptable.

阳茶存之不当,其气速散。阴茶存之不当,其气易变。若黄茶不封,其气渐转而失。此茶气弱化之多由也。

Improper preservation will also reduce the concentration of pneumas. If not preserved in proper ways, the Yaŋ pneumas will be lost soon, and the Yim pneumas will transform easily. For example, when the yellow teas are kept open, the Little Yim pneuma will either flow away or transform into Middle Yaŋ pneuma. Those are the various reasons for the decreasing of the pneuma concentration.

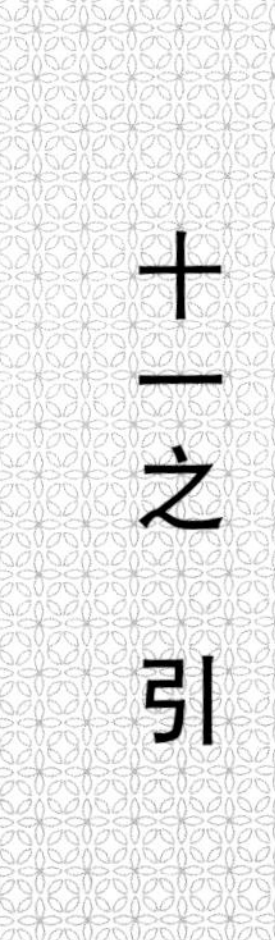

十一之引

Chapter 11 **EXCIPIENTS**

如若茶气略杂或弱,不足健魄,则可配以茶引子而导其为正。若茶气正且佳者,用引则愈佳。陆羽茶经言唐俗有"用葱、姜、枣、橘皮、茱萸、薄荷之等"煮入茶汤似有不妥。此盖指茶引也,非混而入汤,须有应茶之品类、气之上下而引者。

If the pneuma of a tea is slightly impure or weak, and is not good enough to nourish the body souls, excipients can be added to the tea to lead the pneuma to the correct channels. If the pneuma is pure and healthy, the excipients will make the tea even better. The *Tea Scripture* by *Lu Yu* said that in Tang Dynasty, people put scallion (*Allium fistulosum*), ginger (*Zingiber officinale*), jujube (*Drypetes congestiflora*), orange (*Citrus reticulata*) peel, evodia (*Euodia rutaecarpa*), mint (*Mentha haplocalyx*), etc., into the tea soup, which was disliked by *Lu*. Here, these things were used as excipients. They could not be mixed and put into the soup, but be chosen according to the type of tea and the channel the pneumas went into.

绿茶上者以胎菊引,下者以茉莉引;

Into the green teas, we put chrysanthemums (*Chrysanthemum morifolium*) when the pneuma goes to the hand channel, and put jasmine (*Jasminum sambac*) when the pneuma goes to the foot channel.

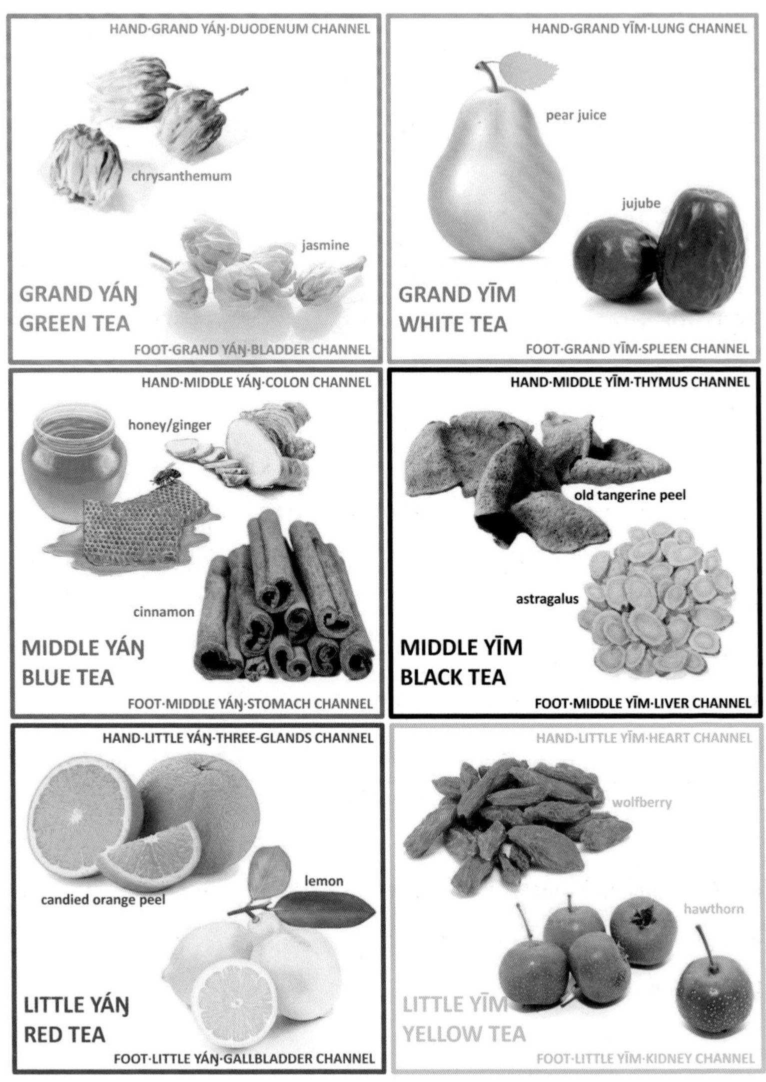

Figure 27. The excipients for the twelve kinds of tea.

青茶上者以蜜姜引,下者以肉桂引;

Into the blue teas, we put the ginger (*Zingiber officinale*) with honey when the pneuma goes to the hand channel, and put cinnamon (*Cinnamomum cassia*) when the pneuma goes to the foot channel.

红茶上者以橙皮引,下者以柠檬葡萄引;

Into the red teas, we put the candied orange (*Citrus sinensis*) peels when the pneuma goes to the hand channel, and put lemons (*Citrus limon*) or grapes (*Vitis vinifera*) when the pneuma goes to the foot channel.

黄茶上者以枸杞引,下者以山楂引;

Into the yellow teas, we put the wolfberries (*Lycium barbarum*) when the pneuma goes to the hand channel, and put hawthorns (*Crataegus pinnatifida*) or highland waterlily (*Nymphaea mexicana*) when the pneuma goes to the foot channel.

黑茶上者以陈皮引,下者以黄芪引;

Into the black teas, we put the old tangerine (*Citrus reticulata* cv. Chachiensis) peels when the pneuma goes to the hand channel, and put astragalus (*Astragalus propinquus*) when the pneuma goes to the foot channel.

白茶上者以梨汁引,下者以红枣引;

Into the white teas, we put the pear (*Pyrus bretschneideri*) juice when the pneuma goes to the hand channel, and put jujubes

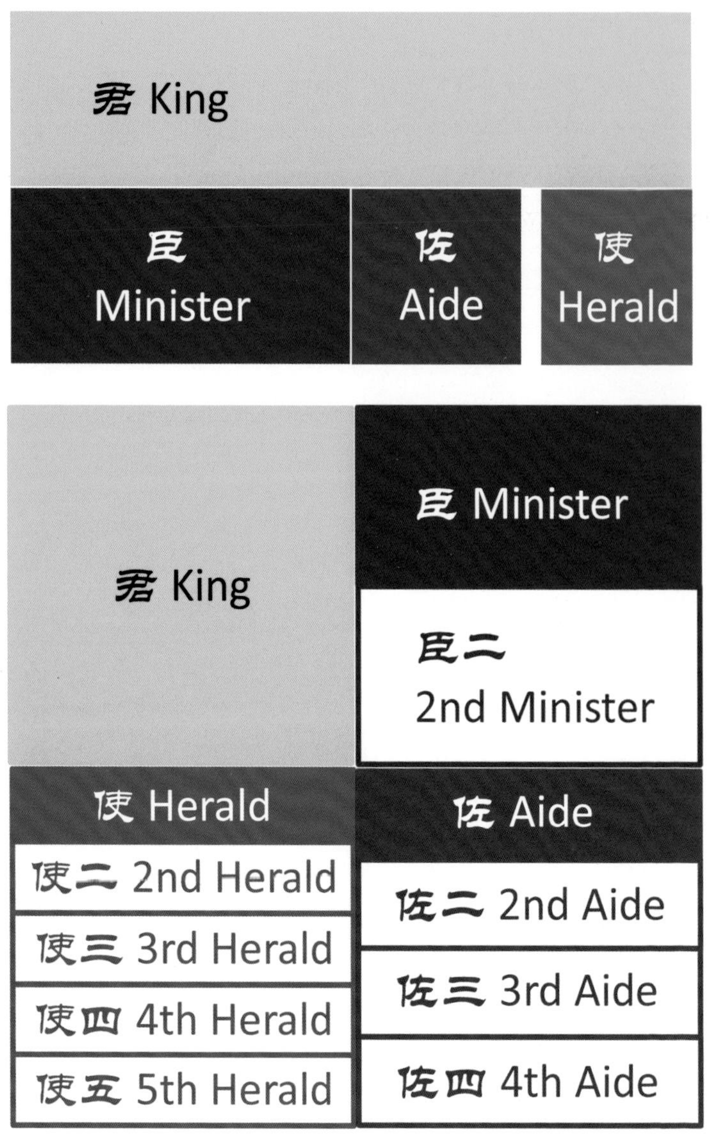

Figure 28. Rule of dosage proportioning according to the Shennong's Canon of Herbs, one King, a half Minister, a fourth Aide, and a fifth Herald.

(*Drypetes congestiflora*) when the pneuma goes to the foot channel.

茶引一二枚即可,多则夺茶之味。

Only one or two pieces of excipients should be put into one soup, or its taste will turn funny.

十二之性

Chapter 12 **ACTIVITY**

茶之性,亦须合于人之性。食药之类,定以三维,曰性、味、归经。归经者,阴阳之属也。味者,甘苦之别也。性者,寒热之分也。

The nature of tea should match the nature of the people. Foods and medicines are defined by three characters, i.e., activity, taste, and meridian going, in Chinese Traditional Medicine. The character of meridian going is the Yim/Yaŋ feature. The character of taste is sweetness, bitterness, acidity, saltiness, and piquancy. The character of activity is rated into five grades from strongly activating the cells to strongly inactivating the cells. To activate the cells is called hotness, while to inactivate the cells is called coldness.

人之有疾,始于营气虚脱,气脉凝滞。饮食男女,少有均和者,是以营气或有不冲。故须考经脉之况而择茶之阴阳,以补营魄。

The illness of the human starts with the weakening of the nutritional pneumas, and the blocking of the meridians. Neither food nor sex is easy to be balanced, and therefore, the nutritional pneumas are seldom full. Thus, one needs to judge the situations of the meridians to choose the proper teas to nourish the body souls.

茶亦可成五味,然则酸苦辣咸非正味也。好茶断无苦涩之理,唯鲜甜爽口者为佳品。苦涩者,人所不喜,多毒害也。良药虽苦,

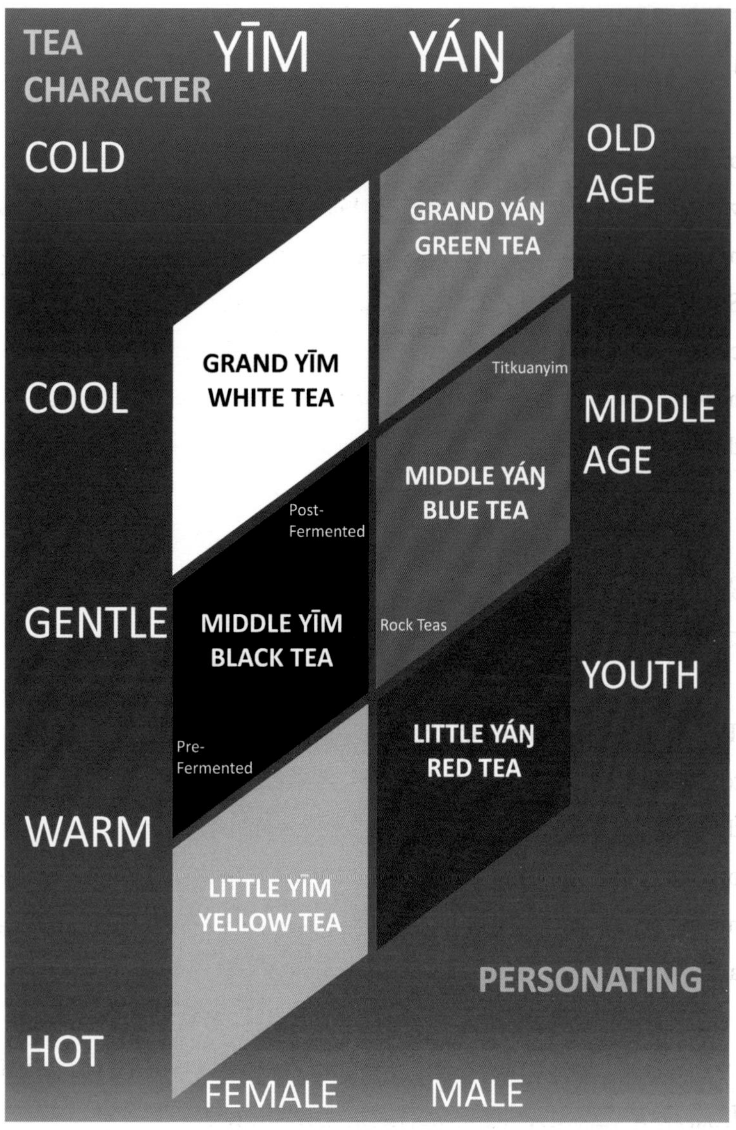

Figure 29. Association between the Yīm/Yáŋ property and activity of the good teas.

用以攻毒,亦非上药,权宜之用也。茶之养生,非救急者,何必苦口,而害其毒焉?

The teas can also be one of the five tastes. However, acidity, bitterness, saltiness, and piquancy are improper for teas. A good tea will never be bitter or astringent, but sweet and smooth. People do not like bitter or astringent food because they are mostly poisonous. Some medicines are bitter and effective for diseases, but cannot be classy medicines and are just temporary expedients. The teas are for health preserving, not for emergency treatment. It's not necessary for the teas to be bitter and poisonous.

人之体质,或寒或热,故宜择茶之寒热以调和之。太阴太阳,茶之老者,体质寒凉;少阴少阳,茶之童者,体质燥热;厥阴阳明,茶之壮者,温凉各异。

Some human bodies may be cold and inactive or hot and active. To balance the activity of the human body, we need to choose the teas according to their activity. Grand Yim and Grand Yaŋ are like the cold or cool old people. Little Yim or Little Yaŋ, are like the hot or warm young people, and Middle Yim or Middle Yaŋ are like the middle-aged people, some are warm, and some are cool. This is the activity principle called "sky cold and earth hot".

故太阴白茶为湿土,如慈母之关爱,虽凉,恰可以解表。表凉里热,体质之顺也,人皆喜之,故白茶无不可饮者。

Grand Yim means an old woman. When drinking white teas, one will feel like staying with a loving mother. Although they are cool, this cool activity can just clean the skin and cure the tetters. Being cool outside and warm inside is the proper situation of the human body, which makes people comfortable. Therefore, the white teas fit all people.

太阳绿茶为寒水,如严父之苛责,极寒,入于内里,殊难承受,惟燥热虚火者宜饮之。

Grand Yaŋ means an old man. When drinking green teas, one will feel like staying with a rigid father. They are cold and go deep into the body, and sometimes make people feel uncomfortable. But those people whose bodies are too active just need to drink the green teas.

少阴黄茶为君火,如少女之怀春,极热,融通心血,周身皆暖,饮之似有返老还童之感,寒湿者犹当用之,如归妹之卦象。

Little Yim means a young girl. When drinking a yellow tea, one will feel like meeting someone who has just fallen into love. It will be very hot, boil the blood, and make the whole body warm. A juvenescence feeling comes up. The yellow teas mostly suit those whose bodies are cold and wet.

少阳红茶为相火,如少男之慕逑,温尔顺人,调谐身心,平和戾气,体质燥寒者宜之。

Little Yaŋ means a little boy. When drinking a red tea, one will feel like meeting someone with all the best wishes. He is nice and makes people happy. Those people whose bodies are dry and cold need to drink more of the red teas.

厥阴黑茶为风木，如中女之温婉，性平略温，宁神静气，宜室宜家，可得永年。

Middle Yim means a middle-aged woman. The black teas are gentle and balanced, and some are slightly warm. They calm people down, make the family harmonious, and extend people's life time.

阳明青茶为燥金，如士人之多性，凉者如铁观音，焙火而有性平至温者如诸岩茶，此茶性如君子之和而不同也，故可视肠胃之寒热中和之。

Middle Yaŋ means a middle-aged man. The blue teas are various in character like the diversity of the gentlemen. Some are cool, such as Titkuanyim, and some are placid and warm, such as the rock teas after having been baked. This character of diversity is called divergent but harmonious. People can choose different blue teas according to the activities of their stomachs and guts to make neutralizations.

故茶者世之罡气也，岂可不问性质，而强饮焉？知阴阳，辨寒热，方可宜茶宜人。

In conclusion, tea contains the best pneumas in the world. One can never drink them without judging their properties. We need to know the Yim or Yaŋ of the pneuma property, and also the coldness or the hotness of the activity, and only then can we choose the right tea for the right person.

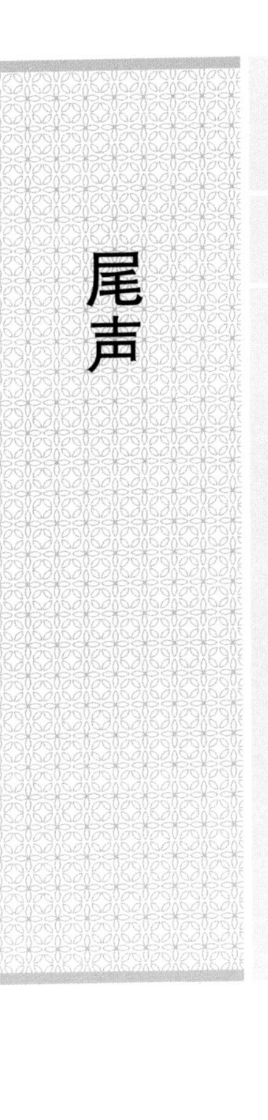

尾声

EPILOGUE

形而上者谓之道,形而下者谓之器。成茶之为物,所用之壶盏,此器也;制茶存茶饮茶之技艺,此术也;茶中所含功能成分,及其理化机理,此法也;茶之阴阳和谐规律,此道也。道可领法,法可化术,术可成器。故知茶道,方能善茶术,而得养于茶之天物也。

In concluding the principles after observations, we approach the Tao (philosophy). When producing the goods after illation, we approach the Qi (material).

The teas, the cups and pots are all materials. The methods of tea production, preservation, and brewing are all techniques. The ingredients of teas and their reactions during all the lives of teas from the farms to the cups are sciences. The principles of the balance between Yim and Yaŋ for the teas are philosophy.

The philosophy leads to sciences. Sciences are translated to techniques. Techniques produce materials.

By getting to know the philosophy and sciences of tea, we can become good at tea techniques, and benefit from the nature of tea.

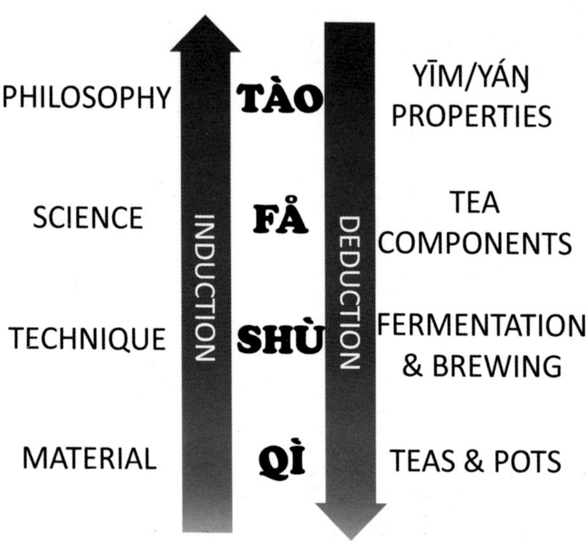

Figure 30. The structure of Yīm/Yáŋ teaism.

Table 6 A sample list of the good teas

Type	Name	Locality	Mountain	Flavor	Reservation	Brewing	Temperature	Activity	Taste	Meridian Channel	Effect
Grand Yaŋ Green Tea	Green Snails	Soochow, Jiangsu	Dongting Hills	chrysan-themum	sealed & frozen	glass	70℃	cold	sweet	Hand · Grand Yaŋ · Duodenum	refreshing
	Monkey Champion	Taiping, Anhui	Yellow Mountain	chrysan-themum	sealed & frozen	glass	70℃	cold	sweet	Hand · Grand Yaŋ · Duodenum	refreshing
	Black Buffalo	Xiangshan, Zhejiang	Dajim Hills	broad bean	sealed & frozen	glass	80℃	cold	sweet	Foot · Grand Yaŋ · Bladder	diuresis
	Dragon Well	Hangzhou, Zhejiasng	Shifeng Hill	broad bean	sealed & frozen	glass	80℃	cold	sweet	Foot · Grand Yaŋ · Bladder	diuresis
Middle Yaŋ Blue Tea	Titkuanyim	Anxi, Fujian	Fengfeng Hills	magnolia	sealed & cellared	covered bowl	90℃	cool	sweet	Hand · Middle Yaŋ · Colon	gut purge
	Oriental Beauty	Xinzhu, Taiwan	Snow Mountain	Sophora honey	sealed & cellared	covered bowl	90℃	cool	sweet	Hand · Middle Yaŋ · Colon	nose relieving
	Red Garment	Chong'an, Fujian	Bohea Mountain	cinnamon	sealed & cellared	covered bowl	92℃	warm	sweet	Foot · Middle Yaŋ · Stomach (Cardia)	stomach calming
	Phoenix Unique	Chao'an, Guangdong	Fenghuang Mountain	osmanthus	sealed & cellared	covered bowl	92℃	gentle	sweet	Foot · Middle Yaŋ · Stomach (Pylorus)	stomach calming

continued

Type	Name	Locality	Mountain	Flavor	Reservation	Brewing	Temperature	Activity	Taste	Meridian Channel	Effect
Little Yaŋ Red Tea	Luoyue Red	Lingyun, Guangxi	Cenwanglao Mountain	Dark chocolate	sealed in earthenware	black pottery	97℃	warm	sweet	Hand · Little Yaŋ · Three-gland	delighting
	Lapsang Souchong	Chong'an, Fujian	Bohea Mountain	white chocolate	sealed in earthenware	black pottery	97℃	warm	sweet	Hand · Little Yaŋ · Three-gland	delighting
	Bonam Red	Yongping, Yunnan	Bonan Mountain	desmodium	sealed in earthenware	black pottery	97℃	warm	sweet	Foot · Little Yaŋ · Gallbladder	cholagogue
	Red Grape	Menghai, Yunnan	Bingmali Mountain	grape	sealed in earthenware	black pottery	97℃	warm	sweet	Foot · Little Yaŋ · Gallbladder	dispelling headache
Little Yim Yellow Tea	Golden Buns	Jiangkou, Guizhou	Fanjing Mountain	wolfberry	sealed in tin can	porcelain pot /boiled	95℃	hot	sweet	Hand · Little Yim · Heart	blood clearing
	Goishi	Kochi, Japan	Yoshino Gawa	eggplant	sealed in tin can	porcelain pot /boiled	95℃	hot	sweet	Hand · Little Yim · Heart	blood clearing
	Luoshan Golden Brick	Zheng'an, Guizhou	Dalou Mountain	water lily	sealed in tin can	boiled	100℃	hot	sweet	Foot · Little Yim · Kidney	kidney clearing
	De'ang Lactics	Mangshi, Yunnan	Laozhong Moutain	cheese	sealed in tin can	steamed	100℃	hot	sweet	Foot · Little Yim · Kidney	kidney clearing

continued

Type	Name	Locality	Mountain	Flavor	Reservation	Brewing	Temperature	Activity	Taste	Meridian Channel	Effect
Middle Yim Black Tea	Tangerine Pu'erh	Lancang, Yunnan	Jingmai Mountain	old tangerine peel	wrapped & hung	red stoneware pot / steamed	99℃	warm	sweet	Hand · Middle Yim · Thymus	nerve calming
	Liubao	Wuzhou, Guangxi	Dagui Mountain	old pomelo peel	wrapped & hung	red stoneware pot / steamed	99℃	warm	sweet	Hand · Middle Yim · Thymus	nerve calming
	Light Pu'erh	Mengla, Yunnan	Yiwu Hills	asiabell	wrapped & hung	red stoneware pot / steamed	99℃	warm	sweet	Foot · Middle Yim · Liver	eyesight improving
	Golden Fungi	Yongzhou, Hunan	Yangming Mountain	astragalus	wrapped & hung	steamed	99℃	gentle	sweet	Foot · Middle Yim · Liver	liver clearing
Grand Yim White Tea	Pear Silver Needles	Jinggu, Yunnan	Wuliang Mountain	pear	sealed & sunbaked	porcelain pot /boiled	100℃	cool	sweet	Hand · Grand Yim · Lung	lung clearing
	White Silver Needles	Fuding, Fujian	Taimu Hills	pear syrup	sealed & sunbaked	porcelain pot /boiled	100℃	cool	sweet	Hand · Grand Yim · Lung	lung clearing

continued

Type	Name	Locality	Mountain	Flavor	Reservation	Brewing	Temperature	Activity	Taste	Meridian Channel	Effect
Grand Yim White Tea	White Peony	Fuding, Fujian	Taimu Hills	smoked jujube	sealed & sunbaked	boiled	100℃	cool	sweet	Foot · Grand Yim · Spleen	immunity enhancing
	Old Eyebrows	Fuding, Fujian	Taimu Hills	jujube	sealed & sunbaked	boiled	100℃	cool	sweet	Foot · Grand Yim · Spleen	immunity enhancing

江苏　苏州　洞庭东山　**East Dongting Hills, Soochow, Jiangsu**

手　太阳　小肠经　绿茶——碧螺春

Hand · Grand Yaŋ · Duodenum: Green Tea, Green Snails

浙江　杭州　狮峰　**Lion Peak, Hangzhou, Zhejiang**

足　太阳　膀胱经　绿茶——龙井

Foot · Grand Yang · Bladder: Green tea, Dragon Well

浙江 永嘉 长夹岭 **Changjia Hills, Yongjia, Zhejiang**

足 太阳 膀胱经 绿茶——乌牛早

Foot · Grand Yaŋ · Bladder: Green Tea, Black Buffalo

福建　安溪　枫凤山　**Fengfeng Hills, Anxi, Fujian**

手　阳明　大肠经　青茶——铁观音

Hand · Middle Yaŋ · Colon: Blue Tea, Titkuanyim

广东　潮安　乌岽山　**Wudong Hills，Chao'an，Guangdong**

足　阳明　胃经(贲门分支)青茶——凤凰单枞

Foot · Middle Yang · Stomach（Cardia branch）：Blue Tea，Phoenix Unique

福建　崇安　武夷山　**Wuyi（Bohea）Mountain，Chong’an，Fujian**

足　阳明　胃经（幽门分支）青茶——大红袍

Foot · Middle Yang · Stomach（Pylorus branch）：Blue Tea，Red Garment

广西 凌云 岑王老山 **Cenwanglao Mountain，Lingyun，Guangxi**

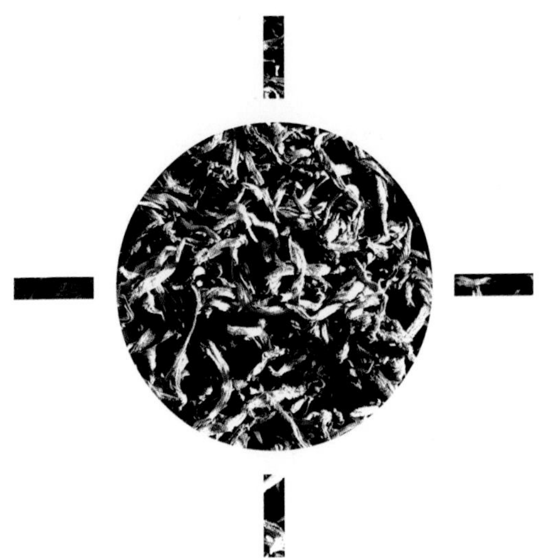

手 少阳 三焦经 红茶——雒越红

Hand · Little Yang · Three-gland：Red Tea，Luoyue Red

福建　南平　桐木关　**Tongmu Pass, Nanping, Fujian**

手　少阳　三焦经　红茶——正山小种

Hand · Little Yang · Three-gland: Red Tea, Lapsang Souchong

云南 永平 博南山 **Bonan Mountain, Yongping, Yunnan**

足 少阳 胆经 红茶——博南红(滇红)
Foot · Little Yang · Gallbladder: Red Tea, Bonan Red Grape

贵州　江口　梵净山　Fanjing Mountain, Jiangkou, Guizhou

手　少阴　心经　黄茶——梵金髻
Hand · Little Yim · Heart: Yellow Tea, Golden Buns

湖南　岳阳　君山　**Junshan Hill, Yueyang, Hunan**

足　少阴　肾经　黄茶——君山金砖

Foot · Little Yim · Kidney: Yellow Tea, Junshan Golden Bricks

贵州　正安　大娄山　**Dalou Mountain, Zheng'an, Guizhou**

足　少阴　肾经　黄茶——正安金锭
Foot · Little Yim · Kidney: Yellow Tea, Zheng'an Golden Bricks

云南 澜沧 景迈山 **Jingmai Mountain, Lancang, Yunnan**

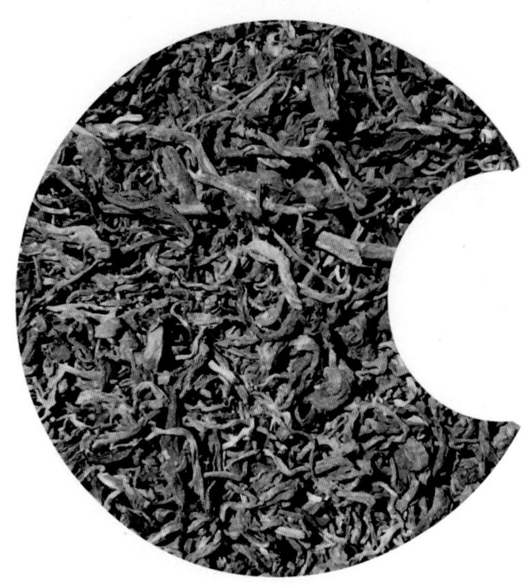

手 厥阴 心包经 黑茶——陈香普洱熟茶
Hand · Middle Yim · Thymus: Black Tea, Tangerine Pu'erh

湖南　永州　阳明山　**Yangming Mountain, Yongzhou, Hunan**

足厥阴肝经黑茶——金花黑茶
Foot · Middle Yim · Liver: Black Tea, Golden Fungi

福建 福鼎 太姥山 **Taimu Hills, Fuding, Fujian**

手太阴肺经白茶——白毫银针

Hand · Grand Yim · Lung: White Tea, White Silver Needle

福建　政和　洞宫山　**Donggong Mountain, Zhenghe, Fujian**

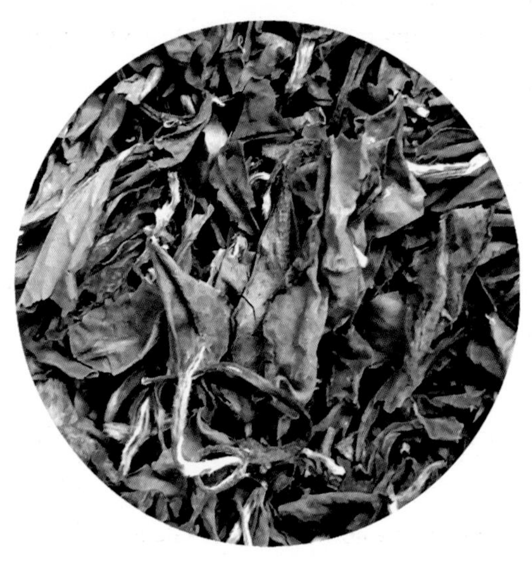

足太阴脾经白茶——寿眉
Foot · Grand Yim · Spleen: White Tea, Old Eyebrows

Author Introduction

LI Hui-Zichen, PhD, Professor of human biology and anthropology at the School of Life Sciences, Fudan University, dean of MOE Key Laboratory of Contemporary Anthropology, vice president of the Institute of Scientific Archaeology, is major in molecular anthropology, focusing on the origin of human populations and civilizations in East Asia. He demonstrated that East Asians originated in two migration waves from Africa. The Chinese Civilization was stimulated by the domestications of rice in the South and millet in the North. Three Sovereigns in early time of Neolithic Age were proved to be ancestors of half of the present Chinese males. The mausoleums were found. He is also an expert of tea science. In 2010, he discovered that six types of tea are related to six major meridians of the human body, and later proved it by experiments. He published more than 270 papers in SCIENCE, NATURE, etc. and several books. Professor LI and his research of historical anthropology was intensively reported by SCIENCE.

The author name on the cover is present as LI H. Zichen. LI is the family name, Hui is the given name, and Zichen is the scholar name. Using scholar name was a prevalent tradition of Chinese people before the establishing of the P. R. China, and is still being kept by part of the scholars today.

	Grand Solar Green Teas	Middle Solar Blue Teas	Little Solar Red Teas
Hand side	Duodenum meridian Monkey Champion 猴魁	Colon meridian Titkuanyim 铁观音	HPT meridian Lapsang Souchong 正山小种
Foot side	Bladder meridian Dragon Well 龙井	Stomach meridian Red Garment 大红袍 Phoenix Unique 凤凰单丛	Gallbladder meridian Yunnan Red Grape 葡红

	Grand Lunar White Teas	Middle Lunar Black Teas	Little Lunar Yellow Teas
Hand side	Lung meridian Silver White Needles 白毫银针	Thymus meridian Tangerine Pu'erh 熟普洱	Heart meridian Golden Buns 梵金髻
Foot side	Spleen meridian Old Eyebrows 寿眉	Liver meridian Golden Fungi 金花茯砖	Kidney meridian Golden Bricks 金锭